SEISMIC METHODS

INSTITUT FRANÇAIS DU PÉTROLE
ÉCOLE NATIONALE SUPÉRIEURE DU PÉTROLE ET DES MOTEURS

Michel LAVERGNE

Professor
Director of the Exploration Graduate Study Center
École Nationale Supérieure du Pétrole et des Moteurs
Institut Français du Pétrole

SEISMIC METHODS

Translation from the French
by Nissim MARSHALL

1989

Graham & Trotman Limited   Sterling House   66 Wilton Road   London

ÉDITIONS TECHNIP   27 RUE GINOUX   75737 PARIS CEDEX 15   **techniP**

Translation of
« Méthodes sismiques »
M. Lavergne
© Éditions Technip, Paris 1986

This Edition, 1989
Graham & Trotman Limited
Sterling House
66 Wilton Road
London

© 1989 Editions Technip, Paris
Softcover reprint of the hardcover 1st edition 1989

ISBN 978-90-481-8243-5

by Imprimerie Nouvelle, 45800 Saint-Jean-de-Braye

preface

This book is a review of the main methods of seismic exploration. It is essentially designed for university students enrolled in Master's degree programs.

For future specialists, and especially for graduate students in petroleum exploration, it is intended only as an introduction to more thorough studies.

This book is also a basic tool for future experts in civil engineering, mining and hydrology, as well as for young seismologists, who are potential users of new geophysical techniques in their different fields.

It is also a useful medium for continuing education and training of non-specialists, to the extent that it outlines essential concepts concerning the acquisition and processing of seismic data.

Michel Lavergne

contents

Preface VII

NOTATIONS 1

Chapter 1

INTRODUCTION 3

Chapter 2

PROPAGATION OF SEISMIC WAVES 7

2.1. **Definitions** 7

2.2. **Review of the theory of elasticity, hypothesis of the homogeneous, isotropic and perfectly elastic medium** 8
    2.2.1. Stress tensor 8
    2.2.2. Strain tensor 9
    2.2.3. Hooke's law 9

2.3. **Wave equations (homogeneous, isotropic, elastic medium)** 11
    2.3.1. Development of wave equations 11
    2.3.2. Solution of the wave equation 13
        2.3.2.1. Point source in homogeneous and isotropic medium 13
        2.3.2.2. Specific case of plane waves 14
        2.3.2.3. General case 16

2.4. **Acoustic impedance** 16

2.5. **Propagation in homogeneous, isotropic and inelastic media** 18

2.6.   Propagation in heterogeneous media   20

2.7   Seismic wave reflection and transmission at the interfaces   23
       2.7.1.   Incident, reflection and transmission angles   23
       2.7.2.   Reflection and transmission coefficients   23
                2.7.2.1.   Reflection and transmission coefficients of potentials   24
                2.7.2.2.   Energy reflection and transmission coefficients   26
                2.7.2.3.   Reflection and transmission coefficients in normal incidence   27

2.8.   Surface waves   30
       2.8.1.   Semi-infinite, homogeneous and isotropic medium, Rayleigh waves   31
       2.8.2.   Semi-infinite, homogeneous and isotropic medium overlaid by a surface layer of
                thickness h. Dispersion   33
                2.8.2.1.   Pseudo-Rayleigh waves   33
                2.8.2.2.   Love waves   35

2.9.   Diffraction   38

Chapter 3
SEISMIC SIGNALS   41

3.1.   Signal and noise. Definitions   41

3.2.   Seismic pulse   42

3.3.   Definitions of Fourier transform, convolution, correlation and auto-
       correlation   42

3.4.   Analytic signal   46

3.5.   Surface noise filtering   47
       3.5.1.   Wavenumber spectrum   47
       3.5.2.   Wavenumber filter   48

3.6.   Reflection seismogram   50
       3.6.1.   Synthetic seismograms   52
                3.6.1.1.   Synthetic seismogram without multiples or transmission losses   53
                3.6.1.2.   Synthetic seismogram with multiples and transmission losses   54
       3.6.2.   Reflection seismograms obtained in seismic prospecting   57
                3.6.2.1.   Short signals   57
                3.6.2.2.   Long sweep signals. Vibroseismic prospecting   57

3.7.   Resolution and detection   60

     3.7.1.   Vertical resolution   60

     3.7.2.   Lateral resolution   62

     3.7.3.   Detection   63

3.8.   The necessity of correct implementation of field surveys and the need for computer processing   63

Chapter 4
REFLECTION SURVEYS   65

4.1.   Data acquisition   65

     4.1.1.   Acquisition systems   65

     4.1.2.   Seismic sources   68

         4.1.2.1.   Land sources   68

         4.1.2.2.   Marine sources   70

         4.1.2.3.   Source directivity   73

     4.1.3.   Seismic detectors   73

         4.1.3.1.   Geophones   74

         4.1.3.2.   Hydrophones   76

     4.1.4.   Digital recording   79

         4.1.4.1.   Digital recording system   79

         4.1.4.2.   Source control   83

         4.1.4.3.   Telemetry recording systems   83

     4.1.5.   Positioning at sea   83

         4.1.5.1.   Radio positioning methods   83

         4.1.5.2.   Satellite positioning methods   86

     4.1.6.   Nature of the signals received in reflection surveying   88

         4.1.6.1.   Horizontal reflector in a homogeneous medium   88

         4.1.6.2.   Dipping reflector in a homogeneous medium   91

         4.1.6.3.   Horizontal tabular stratification   92

     4.1.7.   Three-dimensional surveys   95

     4.1.8.   Shear wave surveys   98

     4.1.9.   Organization of a reflection seismic crew   102

         4.1.9.1.   Land crew   102

         4.1.9.2.   Marine crew   103

     4.1.10.   Implementation cost   105

4.2.   Processing   105

     4.2.1.   Demultiplexing   105

     4.2.2.   Sweep signal correlation   106

4.2.3. Gain recovery and attenuation correction 106

4.2.4. Deconvolution before stacking 108

4.2.5. Static corrections 109
        4.2.5.1. Field statics 109
        4.2.5.2. Residual statics................................... 109

4.2.6. Common midpoint gathering.................................. 111

4.2.7. Velocity analysis 112

4.2.8. NMO correction and CMP 116

4.2.9. Deconvolution after stacking 118
        4.2.9.1. Backus deconvolution 118
        4.2.9.2. Predictive deconvolution 120

4.2.10. Migration 123
        4.2.10.1. Method of summing amplitudes along diffraction curves 123
        4.2.10.2. Wave equation migration 124

4.2.11. Final display of seismic sections 126

4.2.12. Three-dimensional processing 127

4.2.13. Modeling 130

4.3. Conclusions 130

Chapter 5

TRANSMISSION SURVEYS
AND VERTICAL SEISMIC PROFILES 131

5.1. Channel wave transmission..................................... 131

5.2. Velocity tomography by transmission 133

5.3. Vertical seismic profiling 133
      5.3.1. Emission 135
      5.3.2. Detection 137
      5.3.3. Recording 138
      5.3.4. Processing... 141

Chapter 6

REFRACTION SURVEYS...................................... 145

6.1. Data acquisition 147
      6.1.1. Land surveys 147
      6.1.2. Marine surveys 149

6.1.3. Signals received in refraction shooting 151
        6.1.3.1. Horizontal refractor 152
        6.1.3.2. Dipping refractor 153

6.2. Processing and interpretation 155
      6.2.1. Preliminary processing and corrections 155
      6.2.2. Filtering. The tau-p transform 156
      6.2.3. Interpretation: the Gardner method 157

6.3. Importance of refraction surveying 160

Chapter 7
CONCLUSION 161

REFERENCES 163

SUBJECT INDEX 167

notations

A	particle motion amplitude.	j	index or pure imaginary $j = \sqrt{-1}$.
$a(t)$	autocorrelation function.	K	bulk modulus.
a_m	descending wave particle motion amplitude.	k_i	reflection coefficient on interface i.
b_m	ascending wave particle motion amplitude.	$k(t)$	time series of reflection coefficients.
$b(t)$	noise.	k	wavenumber $k = f/c$.
C	phase velocity.	l	$l = \omega/c = 2\pi k$.
C_A	elevation correction.	P	compressional wave.
C_L	Love wave velocity.	p	pressure; slowness $p = 1/V$.
C_R	Rayleigh wave velocity.	psi	pounds/square inch.
c	apparent velocity along interfaces.	VSP	Vertical Seismic Profile.
$c(t)$	convolution function.	OSP	Offset Seismic Profile.
dB	decibel.	Q	Quality-factor.
DP	datum plane (reference plane).	R	real part.
d	geophone spacing.	r	propagation distance.
E	wave intensity.	r_i	sampled desired signal.
e_i	thickness of layer i.	$r(t)$	correlation function.
e_{ij}	strain tensor.	S	shear wave.
e_s	source elevation with respect to datum plane.	SH	polarized shear wave perpendicular to the vertical plane of the profile.
e_R	detector elevation with respect to datum plane.	SV	polarized shear wave in the vertical plane of the profile.
FK	frequency/wavenumber method.	$S(f)$	Fourier transform of $s(t)$.
$F(\omega)$	Fourier transform of signal $f(t)$.	s_i	sampled initial signal.
f	frequency.	$s(t)$	seismic pulse.
f_e	sampling rate.	T	propagation time.
f_i	sampled operator.	t_i	transmission coefficient across interface i.
G	gain.	U	group velocity.
$G(\omega)$	Fourier transform of signal $g(t)$.	$\vec{u}$	particle displacement vector.
h	depth of reflectors.	u	component on Ox of displacement vector.
$h(t)$	subsurface response to a Dirac pulse.	v	component on Oy of displacement vector.
I	imaginary part.		
i	index or subscript.	V	propagation velocity.

$\overline{V}$	RMS velocity.	Λ	wavelength.
V_i	velocity in layer i.	λ	first Lamé parameter.
V_M	average velocity.	μ	second Lamé parameter (shear modulus).
V_p	P-wave velocity.		
V_s	S-wave velocity.	Π	product.
w	component on Oz of displacement vector.	ρ	density.
		Σ	summation.
x	horizontal distance in the profile direction.	σ	Poisson's ratio.
		σ_{ij}	stress tensor.
Y	Young's modulus.	σ_r	reflected stress.
y	horizontal distance perpendicular to the profile direction.	σ_t	transmitted stress.
		τ	time sampling interval; intercept of the Tau-p method.
y_i	sampled seismic trace.		
Z	acoustic impedance.	τ_i	two-way propagation time in layer i.
z	depth.	θ	volume dilatation $\theta = \dfrac{\partial u}{\partial x} + \dfrac{\partial v}{\partial y} + \dfrac{\partial w}{\partial z}$.
α	dip.		
α	absorption coefficient.	θ_i	incidence, reflection, transmission angle.
β	wave propagation direction.		
Δ	Laplacian.	θ_l	critical incidence angle.
Δt	time/distance curve correction.	Φ	dilatational potential.
$\delta(t)$	Dirac pulse.	φ	phase.
δ_{ij}	Kronecker symbol $\delta_{ij} = 1$ if $i = j$ $\phantom{Kronecker symbol \delta_{ij}} = 0$ if $i \neq j$	$\overrightarrow{\Psi}$	distortional potential.
		ω	angular frequency $\omega = 2\pi f$.

1

introduction

Seismic exploration methods consist in creating seismic waves in the subsurface and observing at the surface the waves reflected by the geological beds or refracted along certain interfaces. The disturbances propagate in the subsurface by progressive waves that give rise to reflection and transmission when they reach the boundaries of the geological layers. The reflected waves return to the surface, where they are detected. This is the principle of **seismic reflection** surveys (Fig. 1.1). The transmitted waves may propagate horizontally along a certain distance and then return to the surface. This is the principle of **seismic refraction** surveys (Fig. 1.2).

The geological interfaces identified by reflection prospecting are interfaces of media with different acoustic impedances. By definition, the acoustic impedance is the product of the rock density multiplied by the propagation velocity of the seismic wave. Interfaces observable by seismic refraction prospecting are those in which the propagation velocity in the underlying medium is higher than in the overburden.

Seismic prospecting is an extremely powerful subsurface investigation tool. The depth of penetration of seismic-reflection surveys, for example, is up to 10 km, and its resolution is far better than that of other geophysical methods. On the other hand, it is more expensive than other methods.

Seismic survey

Like any geophysical survey, a seismic survey normally comprises three steps: data acquisition, data processing and interpretation.

(a) **Data acquisition** is achieved by the use of appropriate emission, detection and recording systems in the field. It uses the latest electronics techniques for the digital recording of the data with considerable speed and accuracy.

(b) **Processing** is designed to enhance the signal-to-noise ratio and to enable interpretation of the data. Modern computerized data processing methods have led to the development of increasingly effective software systems.

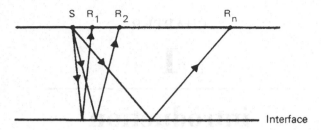

Fig. 1.1 Principle of seismic-reflection surveys. Disturbance at S. Detection at R_1, R_2 .... R_n. Seismic raypaths are shown assuming a homogeneous medium.

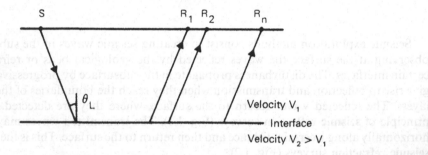

Fig. 1.2 Principle of seismic-refraction surveys. Source at S. Receiver at R_1, R_2 ... R_n. Seismic raypaths are shown assuming a homogeneous medium. θ_L is the critical angle of incidence.

(c) **Interpretation** aims to determine and to characterize the geophysical layers in the subsurface. It is carried out by geologists, and summarizes the overall geological, drilling and geophysical data.

Historical background

The earliest seismic studies date back to the First World War, when the disturbances created by artillery recoil enabled a number of resourceful gunners to locate enemy batteries. The German Mintrop decided to transpose these methods to subsurface exploration, and put the first seismic prospecting into practice around 1920. This led in 1924 to the discovery of the Orchard Salt Dome oilfield in Texas by refraction seismic survey. Soon after, American interpreters detected "reflections" on the refraction recordings and endeavoured to highlight them. This culminated in 1927 in reflection seismic prospecting, which was destined for spectacular growth, and which is the basic method of petroleum exploration today.

1953 witnessed the first magnetic tape recordings, allowing off-line playback and data processing. "Multiple coverage" was invented in 1961. This method consists in increasing the number of emission and detection points, to obtain several reflections at different angles on each subsurface "mirror point". Multiple coverage considerably enhanced the quality of seismic reflection surveys. The advent of digital recording in 1965 vastly broadened the recording "dynamic range", in other words the ratio of the strongest to the weakest signal that can be recorded on magnetic tape. This dynamic range grew from about 40 dB in analog recording (ratio of 100 between the strengths of the strongest and weakest signals) to over 80 dB in digital recording (ratio of 10,000).

Since 1965, advances in seismic data processing have gone hand in hand with computer developments: greater storage capacity, higher data transfer speeds and computation speeds, which were boosted by a factor of around 20 between 1970 and 1980, and continue to advance at the same rate in the 1980s.

Economic importance of seismic prospecting

Total expenditure on applied geophysics in all the market-economy countries amounted to around 3 billion US dollars in 1980. Considered alone, seismic methods account for about 93% of applied geophysical expenditures and 98% of petroleum geophysical expenditures, and these proportions were still valid in 1986. About 95% of the expenditures on geophysical activity in the western world are devoted to petroleum exploration (Whitemore, 1980). This means that it is in the petroleum field that seismic prospecting has found the impetus for its growth.

Seismic exploration also plays a significant role in oceanography and civil engineering. In hydrology and mining exploration, however, its role is relatively smaller than that of more specific and less expensive methods, such as electrical and magnetic methods.

propagation
of seismic waves

2.1 DEFINITIONS

Under the effect of an external force, elastic solids undergo deformations of two types, compression and shear. The theory of elasticity and the basic principle of dynamics show that this leads to the propagation of seismic waves by two different mechanisms, which give rise to two waves propagating independently: compressional waves also called dilatational waves (or longitudinal, or P-waves), and shear waves, also called distortional waves (or transverse, or S-waves). On the passage of the seismic waves, the subsurface particles are set in motion. The surfaces on which the movements are in phase are called **wave surfaces.** In isotropic media, the wave surfaces move perpendicular to themselves.

The interface between the region where the subsurface particles are at rest and the region where they are in motion is by definition the **wavefront.** The normals to the wavefronts in isotropic media are **seismic raypaths** or **rays.** In homogeneous formations, where the seismic velocities are constant, the seismic raypaths are straight lines. When seismic rays pass from one homogeneous formation to another homogeneous formation, the raypaths undergo breaks, as in optics.

While wavefronts have physical significance, raypaths have no real existence. They are introduced for the convenience of argument, in simple cases, but, as a rule, it is better to argue in terms of wavefronts. If a wavefront is known in space and at a given time, its position can be predicted a few moments later by the **Huygens's principle:** any point of a wavefront can be considered as a source of secondary disturbance, and the subsequent wavefronts are the envelopes of all the secondary wavefronts. When the wavefronts reach the subsurface interfaces, they are partly reflected and partly transmitted. If they arrive in oblique incidence, the compressional wave is partly converted into a shear wave and *vice versa.*

In reflection shooting, the raypaths are more or less perpendicular to the geological interfaces, and conversions are slight. The compressional waves are reflected as compressional waves, and the shear waves as shear waves. By contrast, in refraction shooting, in which the incidences are oblique, conversions are substantial, and the compressional waves may give rise to very energetic shear waves.

2.2 REVIEW OF THE THEORY OF ELASTICITY, HYPOTHESIS OF THE HOMOGENEOUS, ISOTROPIC AND PERFECTLY ELASTIC MEDIUM

Under the effect of the stresses generated at the passage of waves, the medium undergoes deformations. After the stresses disappear, it resumes its initial form if its elasticity is perfect. The stresses and strains are distributed throughout the medium and vary in space and time. They are related by equations between the stress tensor and the strain tensor.

2.2.1 Stress tensor

Stress is defined as force per unit area. The stress tensor defines the stresses acting upon each of the six faces of the elementary cube Δx, Δy, Δz (Fig. 2.1).

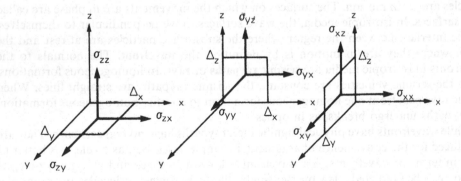

Fig. 2.1 Stress tensor. Stresses on the sides of an elementary cube $\Delta x \, \Delta y \, \Delta z$.

$$\Sigma = \begin{pmatrix} \sigma_{xx} & \sigma_{xy} & \sigma_{xz} \\ \sigma_{yx} & \sigma_{yy} & \sigma_{yz} \\ \sigma_{zx} & \sigma_{zy} & \sigma_{zz} \end{pmatrix} \tag{2.1}$$

The first subscript indicates the facet of the cube, and the second the direction of the stress.

2.2.2 Strain tensor

The strain tensor represents the strains undergone by the elementary cube under the effect of stresses. Let $\vec{u}$ (components u, v, w) be the displacement of a point M with coordinates (x, y, z). The relative change in shape of the cube is defined by the strain tensor:

$$E = \begin{pmatrix} e_{xx} & e_{xy} & e_{xz} \\ e_{yx} & e_{yy} & e_{yz} \\ e_{zx} & e_{zy} & e_{zz} \end{pmatrix} \tag{2.2}$$

The strains e_{ij} are expressed from the displacements u_i by the expression:

$$e_{ij} = \frac{1}{2}\left(\frac{\partial u_i}{\partial x_j} + \frac{\partial u_j}{\partial x_i}\right)$$

in other words:

$$\left|\begin{array}{l} e_{xx} = \dfrac{\partial u}{\partial x};\ e_{yy} = \dfrac{\partial v}{\partial y};\ e_{zz} = \dfrac{\partial w}{\partial z} \quad \text{(compressional terms)} \\[2mm] e_{xz} = e_{yx} = \dfrac{1}{2}\left(\dfrac{\partial v}{\partial x} + \dfrac{\partial u}{\partial y}\right) \\[2mm] e_{yz} = e_{zy} = \dfrac{1}{2}\left(\dfrac{\partial w}{\partial y} + \dfrac{\partial v}{\partial z}\right) \\[2mm] e_{zx} = e_{xz} = \dfrac{1}{2}\left(\dfrac{\partial u}{\partial z} + \dfrac{\partial w}{\partial x}\right) \end{array}\right\}\quad \text{(shear terms)} \tag{2.3}$$

The stress tensor and strain tensor are generally symmetrical, with:

$$\sigma_{ij} = \sigma_{ji}$$
$$e_{ij} = e_{ji}$$

2.2.3 Hooke's law

Hooke's law expresses the linear relationships existing between strains and stresses, when strains are small. In **anisotropic media,** the relationships between stresses and strains depend on 21 elastic parameters. **In transversally isotropic media** (identical elastic properties in two perpendicular directions to an axis of symmetry), they depend on only five parameters. **In homogeneous and isotropic media,** the relationships between stresses and strains depend on only two parameters.

Hooke's law is therefore written:

$$\sigma_{ij} = \lambda\theta\delta_{ij} + 2\mu e_{ij} \tag{2.4}$$

where δ_{ij} is the Kronecker symbol ($\delta_{ij} = 1$ if $i = j$, and 0 if $i \neq j$) and λ and μ are the Lamé constants.

$$\theta = \frac{\partial u}{\partial x} + \frac{\partial v}{\partial y} + \frac{\partial w}{\partial z} = \text{div} . \vec{u} \tag{2.5}$$

is the volume dilatation.

By developing (2.4), six linear equations are obtained between stresses and displacements:

$$\sigma_{xx} = \lambda\theta + 2\mu\frac{\partial u}{\partial x}$$

$$\sigma_{yy} = \lambda\theta + 2\mu\frac{\partial v}{\partial y}$$

$$\sigma_{zz} = \lambda\theta + 2\mu\frac{\partial w}{\partial z}$$

$$\sigma_{xy} = \mu\left(\frac{\partial v}{\partial x} + \frac{\partial u}{\partial y}\right)$$ (2.6)

$$\sigma_{yz} = \mu\left(\frac{\partial w}{\partial y} + \frac{\partial v}{\partial z}\right)$$

$$\sigma_{zx} = \mu\left(\frac{\partial u}{\partial z} + \frac{\partial w}{\partial x}\right)$$

Homogeneous and isotropic media are defined by two elastic parameters selected among the following:

(a) **Lamé parameter** λ.

(b) **Lamé parameter** μ, also called the **shear or rigidity modulus,** which measures the ratio of the tangential stress to the corresponding shear:

$$\mu = \frac{1}{2}\frac{\sigma_{xy}}{e_{xy}}$$

(c) **Young's modulus** Y:

$$Y = \frac{\sigma_{xx}}{e_{xx}}$$

ratio of the normal stress to the corresponding compression. It can be shown that:

$$Y = \frac{\mu(3\lambda + 2\mu)}{\lambda + \mu}$$ (2.7)

(d) **Bulk modulus** K:

$$K = -\frac{\Delta p}{\theta}$$

the ratio of the pressure variation to the volume dilatation.

It can be shown that:

$$K = \lambda + \frac{2}{3}\mu$$ (2.8)

(e) **Poisson's ratio** σ:

$$\sigma = -\frac{e_{yy}}{e_{xx}}$$

the ratio of compression in one direction to extension in the perpendicular direction. It can be shown that:

$$\sigma = \frac{\lambda}{2(\lambda + \mu)} \tag{2.9}$$

The order of magnitude of the parameters λ, μ, Y and K generally ranges between 10^9 and 10^{11} N/m^2 for commonly-encountered formations. σ varies from 0 to 1/2 in isotropic media.

2.3 WAVE EQUATIONS
(homogeneous, isotropic, elastic medium)

2.3.1 Development of wave equations

To obtain wave equations, the **basic principle of dynamics** is applied to the elementary cube defined in Section 2.2, by writing that the sum of the components applied in a given direction, on the six sides of the cube of unit volume, is equal to the product of the density multiplied by the acceleration.

In the Ox direction, for example, the basic principle of dynamics is written:

$$\frac{\partial \sigma_{xx}}{\partial x} + \frac{\partial \sigma_{yx}}{\partial y} + \frac{\partial \sigma_{zx}}{\partial z} = \rho \frac{\partial^2 u}{\partial t^2} \tag{2.10}$$

Let us replace the stresses by strains by applying Hooke's law. This gives the wave equation in the Ox direction:

$$\lambda \frac{\partial \theta}{\partial x} + 2\mu \left(\frac{\partial e_{xx}}{\partial x} + \frac{\partial e_{yx}}{\partial y} + \frac{\partial e_{zx}}{\partial z} \right) = \rho \frac{\partial^2 u}{\partial t^2}$$

namely:

$$(\lambda + \mu) \frac{\partial \theta}{\partial x} + \mu \Delta u = \rho \frac{\partial^2 u}{\partial t^2} \tag{2.11}$$

$$\Delta u = \frac{\partial^2 u}{\partial x^2} + \frac{\partial^2 u}{\partial y^2} + \frac{\partial^2 u}{\partial z^2}$$

is the **Laplacian of u.**

Similar expressions are obtained in the Oy and Oz directions. The wave equation can hence be written in the general form:

$$(\lambda + \mu) \overrightarrow{\text{grad}} \, (\text{div} \cdot \vec{u}) + \mu \Delta \vec{u} = \rho \frac{\partial^2 \vec{u}}{\partial t^2} \tag{2.12}$$

where $\vec{u}$ is the displacement of point M at the passage of the wave.

The **Helmholtz method** can be used to separate the compressional and shear waves. This consists in breaking down the displacement vector into a dilatational component and a distortional component, in the form:

$$\vec{u} = \overrightarrow{\text{grad }} \Phi + \overrightarrow{\text{rot }} \vec{\Psi} \qquad (2.13)$$

where

Φ = dilatational potential (scalar potential),
$\vec{\Psi}$ = distortional potential (vector potential, coordinates Ψ_1, Ψ_2, Ψ_3).

This gives:

$$\left| \begin{array}{l} u = \dfrac{\partial \Phi}{\partial x} + \dfrac{\partial \Psi_3}{\partial y} - \dfrac{\partial \Psi_2}{\partial z} \\[2mm] v = \dfrac{\partial \Phi}{\partial y} + \dfrac{\partial \Psi_1}{\partial z} - \dfrac{\partial \Psi_3}{\partial x} \\[2mm] w = \dfrac{\partial \Phi}{\partial z} + \dfrac{\partial \Psi_2}{\partial x} - \dfrac{\partial \Psi_1}{\partial y} \end{array} \right. \qquad (2.14)$$

The wave equation can be seen to be satisfied if Φ and $\vec{\Psi}$ are solutions of the equations:

$$\Delta \Phi = \frac{1}{V_p^2} \frac{\partial^2 \Phi}{\partial t^2} \quad \text{with} \quad V_p = \sqrt{\frac{\lambda + 2\mu}{\rho}}$$

$$\Delta \Psi_i = \frac{1}{V_s^2} \frac{\partial^2 \Psi_i}{\partial t^2} \quad \text{with} \quad V_s = \sqrt{\frac{\mu}{\rho}} \qquad (2.15)$$

$$\text{and} \quad V_i \in (1, 2, 3)$$

The first equation corresponds to the propagation of a compressional wave at velocity:

$$V_p = \sqrt{\frac{\lambda + 2\mu}{\rho}}$$

The second group of equations corresponds to the propagation of a shear wave at velocity:

$$V_s = \sqrt{\frac{\mu}{\rho}}$$

Note that:

V_s is always lower than V_p,
V_s is zero if the rigidity modulus is zero.

Shear waves do not propagate in a liquid medium.

The orders of magnitude of the compressional and shear wave velocities and of the densities of commonly-found formations are given in Table 2.1.

Table 2.1

Orders of magnitude of *P*- and *S*-wave propagation velocities and rock densities

Type of formation	*P*-wave velocity (m/s)	*S*-wave velocity (m/s)	Density
Weathered rocks	300- 700	100- 300	1.7-2.4
Dry sands	400-1200	100- 500	1.5-1.7
Wet sands	1500-4000	400-1200	1.9-2.1
Clays	1100-2500	200- 800	2.0-2.4
Marls	2000-3000	750-1500	2.1-2.6
Sandstones	3000-4500	1200-2800	2.1-2.4
Limestones	3500-6000	2000-3300	2.4-2.7
Chalk	2300-2600	1100-1300	1.8-2.3
Salt	4500-5500	2500-3100	2.1-2.3
Anhydrite	4000-5500	2200-3100	2.9-3
Dolomite	3500-6500	1900-3600	2.5-2.9
Granite	4500-6000	2500-3300	2.5-2.7
Basalt	5000-6000	2800-3400	2.7-3.1
Coal	2200-2700	1000-1400	1.3-1.8
Water	1450-1500	—	1
Ice	3400-3800	1700-1900	0.9
Oil	1200-1250	—	0.6-0.9

2.3.2 Solution of the wave equation

2.3.2.1 Point source in homogeneous and isotropic medium

Assuming a symmetrical disturbance at the origin, only the compressional wave is generated, and the wave equation is expressed in spherical coordinates in the form:

$$\frac{\partial^2 \Phi}{\partial r^2} + \frac{2}{r}\frac{\partial \Phi}{\partial r} = \frac{1}{V_p^2}\frac{\partial^2 \Phi}{\partial t^2} \tag{2.16}$$

where r represents the distance from the source and t the propagation time.

The solution is written in the form:

$$\Phi = \frac{1}{r} s\left(t - \frac{r}{V_p}\right) \tag{2.17}$$

where $s(t)$ represents the disturbance function.

The solution represents a spherical wave propagating symmetrically from the origin and attenuated in proportion to the distance traveled r. The wavefronts are spheres, and particle motion is parallel to the propagation direction and perpendicular to the wavefronts.

Assuming an asymmetric disturbance, a shear wave may be generated. The wavefronts are also spheres: it can be shown that particle motion is now perpendicular to the propagation direction and parallel to the wavefronts.

2.3.2.2 Specific case of plane waves

Let us consider plane waves perpendicular to the vertical plane (Fig. 2.2). The solutions of the propagation equation can be obtained simply:

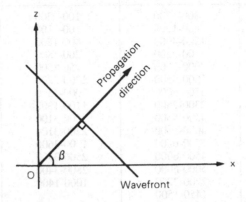

Fig. 2.2   Wavefront and propagation direction for a plane wave perpendicular to plane xz. Homogeneous and isotropic medium.

- **For compressional waves:**

By writing that the particle motion is contained in the vertical plane and is in the propagation direction:

$$v \equiv 0$$
$$u, w \text{ independent of } y$$

Propagation is defined by the dilatational potential Φ and the wave equation is written:

$$\frac{\partial^2 \Phi}{\partial x^2} + \frac{\partial^2 \Phi}{\partial z^2} = \frac{1}{V_{p_2}} \frac{\partial^2 \Phi}{\partial t^2} \tag{2.18}$$

Its solution is:

$$\Phi = s\left(t - \frac{x}{c_p} - \frac{z}{c_p'}\right) \tag{2.19}$$

where

$$\left| \begin{array}{l} c_p = \dfrac{V_p}{\cos \beta} \\[2mm] c_p' = \dfrac{V_p}{\sin \beta} \end{array} \right. \tag{2.20}$$

denote the apparent propagation velocities of the compressional wave along x and z, where β is the propagation direction. $s(t)$ is the disturbance function.

Particle motion is given by:

$$\left|\begin{array}{l} u = \dfrac{\partial \Phi}{\partial x} \\[2ex] w = \dfrac{\partial \Phi}{\partial z} \end{array}\right. \tag{2.21}$$

- **For shear waves:**

Particle motion is perpendicular to the propagation direction. Two types of shear wave can be distinguished: **SV-waves** for which the particle motion is contained in the vertical plane of the source geophone system, and **SH-waves** for which the particle motion is perpendicular to the vertical plane of the system.

(a) *SV*-waves

It is stated that the particle motion is contained in the **vertical plane** and is perpendicular to the propagation direction:

$$\left|\begin{array}{l} v \equiv 0 \\[1ex] u,\ w \text{ independent of } y \end{array}\right.$$

Propagation is defined by the component Ψ_2 of the distortional potential and the wave equation is written:

$$\frac{\partial^2 \Psi_2}{\partial x^2} + \frac{\partial^2 \Psi_2}{\partial z^2} = \frac{1}{V_{s^2}} \frac{\partial^2 \Psi_2}{\partial t^2} \tag{2.22}$$

Its solution is:

$$\Psi_2 = s_2 \left(t - \frac{x}{c_s} - \frac{z}{c_s'} \right) \tag{2.23}$$

where

$$\left|\begin{array}{l} c_s = \dfrac{V_s}{\cos \beta} \\[2ex] c_s' = \dfrac{V_s}{\sin \beta} \end{array}\right. \tag{2.24}$$

denote the apparent propagation velocities of the shear wave along x and z; $s_2(t)$ is the disturbance function.

Particle motion is given by:

$$\left|\begin{array}{l} u = -\dfrac{\partial \Psi_2}{\partial z} \\[2ex] w = \dfrac{\partial \Psi_2}{\partial x} \end{array}\right. \tag{2.25}$$

(b) *SH*-waves

It is stated that the motion is contained in the horizontal plane and is perpendicular to the propagation direction:

$$u \equiv 0$$
$$w \equiv 0$$
$$v \text{ independent of } y$$

Propagation is defined by the components Ψ_1 and Ψ_3 of the distortional potential with the same wave equation as (2.22) and similar solutions.

Particle motion is given by:

$$v = \frac{\partial \Psi_1}{\partial z} - \frac{\partial \Psi_3}{\partial x} \tag{2.26}$$

2.3.2.3 General case

The solution of the wave equation is much more complicated in the general case. The problem can be solved numerically by using processes for solving equations with partial derivatives by the finite difference method, or by other more elaborate methods. These techniques are described in specialized works, and particularly in the book by Jon Claerbout (1976).

2.4 ACOUSTIC IMPEDANCE

By definition, the acoustic impedance of a rock is the product of its density ρ multiplied by the seismic velocity V_p or V_s. Thus an acoustic impedance ρV_p exists for compressional waves, and an acoustic impedance ρV_s for shear waves. It can be shown that the acoustic impedance thus defined is equal to the ratio of the stress to the particle velocity, for plane waves in homogeneous media.

Let us confirm the specific case of plane waves (compressional waves or shear waves) propagating in a direction parallel to the Oz axis.

For **compressional waves**, particle motion is along Oz and is given by a solution of the form:

$$w = g\left(t - \frac{z}{V_p}\right) \tag{2.27}$$

We also have:

$$\frac{\partial w}{\partial z} = \frac{-1}{V_p} \frac{\partial w}{\partial t} \tag{2.28}$$

Hooke's law is written:

$$\sigma_{zz} = (\lambda + 2\mu) e_{zz}$$
$$= (\lambda + 2\mu) \frac{\partial w}{\partial z}$$
$$= -\frac{\lambda + 2\mu}{V_p} \frac{\partial w}{\partial t} \tag{2.29}$$

and by noting that $\lambda + 2\mu = \rho V_p^2$:

$$\sigma_{zz} = -\rho V_p \frac{\partial w}{\partial t} \qquad (2.30)$$

The compressional wave acoustic impedance is hence written:

$$\rho V_p = -\frac{\sigma_{zz}}{\dfrac{\partial w}{\partial t}} \qquad (2.31)$$

$$\boxed{\begin{array}{ll} \text{Acoustic impedance} \\ \text{of } P\text{-waves} \end{array} = -\frac{\text{Dilatational stress}}{\text{Particle velocity}}} \qquad (2.32)$$

For **shear waves**, assuming that particle motion is along Ox, particle displacement is given by a solution of the form:

$$u = h\left(t - \frac{z}{V_s}\right) \qquad (2.33)$$

and we have:

$$\frac{\partial u}{\partial z} = -\frac{1}{V_s}\frac{\partial u}{\partial t} \qquad (2.34)$$

Hooke's law is written:

$$\begin{aligned}
\sigma_{zx} &= 2\mu e_{zx} \\
&= \mu \frac{\partial u}{\partial z} \\
&= -\frac{\mu}{V_s}\frac{\partial u}{\partial t} \qquad (2.35)
\end{aligned}$$

and by noting that $\mu = \rho V_s^2$:

$$\sigma_{zx} = -\rho V_s \frac{\partial u}{\partial t} \qquad (2.36)$$

The shear wave acoustic impedance is hence written:

$$\rho V_s = -\frac{\sigma_{zx}}{\dfrac{\partial u}{\partial t}} \qquad (2.37)$$

$$\boxed{\begin{array}{ll} \text{Acoustic impedance} \\ \text{of } S\text{-waves} \end{array} = -\frac{\text{Distortional stress}}{\text{Particle velocity}}} \qquad (2.38)$$

2.5 PROPAGATION IN HOMOGENEOUS, ISOTROPIC AND INELASTIC MEDIA

If the medium is not perfectly elastic, which is often the case in the subsurface, the seismic wave undergoes some dissipation, part of the seismic energy being converted irreversibly into heat. This process is called **absorption.** Absorption is related to the frequency of the seismic waves and, as a rule, the higher the frequency, the greater the absorption.

Let us consider a plane wave propagating at velocity V in a homogeneous, isotropic and elastic medium, in the positive z direction. The movement of a harmonic component with angular frequency $\omega = 2\pi f$ can be written:

$$A = A_0 \exp\left(j\omega\left(t - \frac{z}{v}\right)\right) \tag{2.39}$$

If the medium is homogeneous, isotropic and inelastic, the motion is written:

$$A = A_0 \exp\left(j\omega\left(t - \frac{z}{v}\right)\right) \exp\left(-\alpha z\right) \tag{2.40}$$

where α is by definition the **absorption coefficient.**

Measurements taken in several university laboratories and at the IFP have shown that, in the frequency range concerned in seismic exploration (between 10 and 250 Hz, for example), the absorption coefficient α can be considered as a first approximation as proportional to the frequency. We can therefore write:

$$\alpha \cong \frac{\pi f}{QV} \tag{2.41}$$

where
  f = frequency,
  V = propagation,
  Q = Quality-factor.

The Quality-factor is independent of the frequency.

The process is often expressed in terms of the dissipation factor Q^{-1}, which characterizes the absorbent properties of the subsurface. The logarithmic decrement $\Delta = \pi/Q$ is the attenuation per cycle.

Absorption can be expressed in nepers/cycle or nepers/wavelength, but is usually expressed in decibels/wavelength (1 Np = 20 $\log_{10} e$ = 8.686 dB). For compressional waves, absorption is generally as follows:

(a) About 0.05 dB/wavelength in compact formations (compact limestones, granite).
(b) About 0.2 dB/wavelength in clays, marls and sands.
(c) About 0.5 dB/wavelength in loose formations.

The Quality-factor generally ranges between 30 and 600 (Q = 30 for viscoelastic formations, Q = 600 for elastic formations). A number of values of the Quality-factor for commonly-found formations are given in Table 2.2 for compressional waves. For shear waves, Q is often slightly lower, especially in water-saturated formations.

Table 2.2

Orders of magnitude of Q-factors of rocks (P-waves)

Type of formation	Q-factor
Clays and marls	30- 70
Sands and sandstones	70-150
Limestones and dolomites	100-600
Granites and basalts..............................	200-600

a. Complex velocity

In certain cases, and especially if Q is not too small ($Q > 20$), the absorption term can be introduced conveniently into the propagation, by using a complex propagation velocity.

In fact, the motion of the harmonic component $\omega = 2\pi f$ can be written:

$$A = A_0 \exp\left(j\omega\left(t - \frac{z}{V}\right)\right) \exp -\left(\frac{\pi f}{V} Q^{-1}z\right)$$

$$= A_0 \exp\left(j\omega\left(t - \frac{z}{V} - \frac{Q^{-1}}{2j}\frac{z}{V}\right)\right) \tag{2.42}$$

where z/V is the propagation term (real number) and

$$\frac{Q^{-1}}{2j}\frac{z}{V}$$

is the dissipation term (pure imaginary number).

The motion of the harmonic component ω in an absorbent medium is hence written:

$$A = A_0 \exp\left(j\omega\left[t - \frac{z}{V}\left(1 - \frac{jQ^{-1}}{2}\right)\right]\right) \tag{2.43}$$

and by setting:

$$\frac{1}{V}\left(1 - \frac{jQ^{-1}}{2}\right) = \frac{1}{V'} \tag{2.44}$$

it is simply written:

$$A = A_0 \exp\left(j\omega\left(t - \frac{z}{V'}\right)\right) \tag{2.45}$$

where the complex velocity V' has replaced, in absorbent medium, the real propagation velocity V in the elastic medium.

b. *Order of magnitude of real and imaginary components*
   of the complex velocity

$$V' = V \frac{1}{1 - \dfrac{jQ^{-1}}{2}} = V \frac{1 + \dfrac{jQ^{-1}}{2}}{1 + \dfrac{Q^{-2}}{4}}$$

The Quality-factor is about 100 (ranging between 30 and 600). Hence $Q^{-2}/4$ is negligible in comparison with 1, and we can write:

$$V' \cong \left(1 + j \frac{Q^{-1}}{2}\right) V \qquad (2.46)$$

Thus, as a first approximation, the absorption can be introduced by replacing the real propagation velocity V by a complex velocity V', which includes a real part equal to V and a pure imaginary part proportional to the dissipation factor Q^{-1}. The dissipation term is generally much smaller than the propagation term, 60 to 1200 times smaller depending on the formation[1].

2.6 PROPAGATION IN HETEROGENEOUS MEDIA

In heterogeneous media, the elastic parameters and the density may vary continuously or discontinuously. As a first approximation, it can be considered that seismic wave propagation is governed, as in optics, by the Fermat principle, which states that seismic raypaths follow paths of minimum traveltime. Thus laws exist similar to Snell's laws, which relate the incident angle to the reflection and transmission angles on the interfaces: the reflection angle is equal to the incident angle, and the transmission angle is greater or smaller than the incident angle depending on whether the seismic velocity is increased or decreased. As in optics, particularly for a velocity increase, a critical incident angle θ_l exists at which the transmission angle is 90°. The transmitted ray propagates along the interface and sends energy to the upper medium at all times. This is the principle of refraction shooting.

However, propagation laws are more complicated in seismic exploration than in optics, due to the presence of two types of propagation: in compressional waves and shear waves.

In sedimentary formations, it often happens that the increased compaction with depth causes a progressive increase in seismic velocities (Dobrin, 1976). The seismic raypaths are then unbroken lines, whose curvature increases directly with the velocity gradient (Fig. 2.3).

---

(1) If dissipation occurs, the propagation velocity is also dispersive, i.e. it depends on the frequency. However, dispersion is practically negligible in the seismic frequency band, if Q is higher than 20.

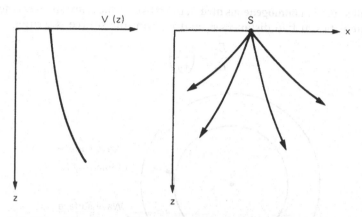

Fig. 2.3 Seismic raypaths in a medium with velocity increasing with depth.

The porosity and the mineralogical composition of the grains also play an important role in the establishment of seismic velocities. Thus, in sandstones partly saturated with water, the velocity is often obtained from the fluid velocity and from the velocity of the rock matrix by the formula (Wyllie *et al.*, 1958):

$$\frac{1}{V} = \frac{\Phi}{V_F} + \frac{1 - \Phi}{V_M} \tag{2.47}$$

where
    V = velocity of compressional waves in the rock,
    V_F = velocity in the pore fluid,
    V_M = velocity in the rock matrix,
    Φ = porosity.

Geometric spreading

Even in homogeneous and elastic media (no absorption), particle motion is generally attenuated during propagation, due to the expansion of the wavefronts. This is geometric spreading, or attenuation by spreading. At a given point, for a harmonic component with angular frequency ω, the particle motion can be written in elastic media:

$$A = A_0 \exp(j\omega t) \tag{2.48}$$

The energy per unit volume, or intensity, is:

$$E = \frac{1}{2} \rho \omega^2 A_0^2 \tag{2.49}$$

where ρ denotes the rock density and A_0 the maximum particle displacement.

Geometric spreading does not cause any loss of energy. For a point source, for example, the same quantity of energy is spread over increasingly wide wave surfaces, and the energy per unit volume decreases as a function of the distance of the wave from the source.

With a point source in **homogeneous media,** the wave surfaces are spheres (Fig. 2.4) and the ratio of intensities at two distances r_1 and r_2 from the source is written:

$$\frac{E_2}{E_1} = \left(\frac{r_1}{r_2}\right)^2 \qquad\qquad (2.50)$$

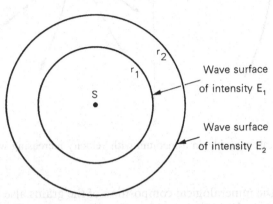

Fig. 2.4   Wave surface for a point source in a homogeneous medium.

The ratio of particle motion is therefore:

$$\frac{A_2}{A_1} = \frac{r_1}{r_2} \qquad\qquad (2.51)$$

and the attenuation of particle motion due to geometric spreading is proportional to the distance traveled by the wave. The amplitude is attenuated according to an inverse law $1/r$ (or $1/Vt$, where t is the propagation time and V the velocity).

In **heterogeneous media,** the wave surfaces are not spherical, and the ratio of intensities at distances r_1 and r_2 from the source is no longer $\left(\dfrac{r_1}{r_2}\right)^2$.

In stratified media with propagation velocity increasing with depth, it can be shown (Newman, 1973) that the attenuation of particle motion due to geometric spreading takes place according to a law:

$$\frac{1}{T}\frac{V_1}{\overline{V}^2(T)}$$

where

T = propagation time,

V_1 = velocity in the upper medium,

$\overline{V}(T)$ = RMS velocity at time $T^{(2)}$.

This formula is valid for a point source located close to the surface. It can theoretically be used to make the geometric spreading correction in reflection surveying.

---

(2) The Root Mean Square (RMS) velocity $\overline{V}$ defined by:

$$\overline{V}^2 = \frac{1}{T_n}\int_0^{T_n} V^2 \, dT$$

is discussed in Section 4.1.6.

2.7 SEISMIC WAVE REFLECTION AND TRANSMISSION AT THE INTERFACES

When the wavefronts reach the subsurface interfaces, they are partly reflected and partly transmitted, in accordance with the boundary conditions of elastic media: continuity of stresses and displacements at the passage of the interfaces. To analyze reflection and transmission, we shall restrict ourselves, for the sake of simplicity, to plane waves incident on plane and horizontal interfaces of infinite size.

Let us consider three possible types of incident wave:

(a) Compressional wave (P-wave): particle motion perpendicular to the wavefront and lying in the vertical propagation plane.
(b) Type SV-shear wave: particle motion parallel to the wavefront and lying in the vertical propagation plane.
(c) Type SH-shear wave: particle motion perpendicular to the vertical propagation plane.

The first two cases lead to conversions of P-waves into SV waves and *vice versa*. The third case does not give rise to conversion, in so far as the interfaces are parallel to the particle motion; the SH-type incident waves are only reflected and transmitted as SH-waves.

Hence two major families are distinguished:

(a) The family of P- and SV-waves, subject to conversions.
(b) The family of SH-waves, without conversion, as a first approximation.

2.7.1 Incident, reflection and transmission angles

The **angular relationships** between the incident, reflected and transmitted raypaths in the different types of wave are determined from Huygens's principle and lead to the generalized Snell's law (Fig. 2.5):

$$\frac{\sin \theta_1}{V_{p_1}} = \frac{\sin \theta_1'}{V_{s_1}} = \frac{\sin \theta_2}{V_{p_2}} = \frac{\sin \theta_2'}{V_{s_2}} \tag{2.52}$$

where $\theta_1, \theta_1', \theta_2$ and θ_2' represent the incident, reflection and transmission angles across the interface (see Fig. 2.5), and V_{p_1}, V_{s_1}, V_{p_2} and V_{s_2} are the propagation velocities of the compressional and shear waves in media 1 and 2 respectively.

2.7.2 Reflection and transmission coefficients

The energy distribution between the incident, reflected and transmitted waves can be calculated by introducing the boundary conditions on the displacements and stresses, on either side of the interface (Ewing *et al.*, 1957).

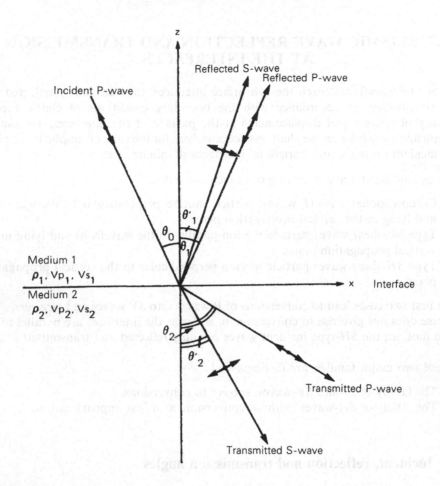

Fig. 2.5 Angular relationships between incident, reflected and transmitted rays for different wave types. Case of a plane incident compressional wave whose front is perpendicular to the plane of the figure. Particle motion is shown schematically.

2.7.2.1 Reflection and transmission coefficients of potentials

First case: incident P- and SV-waves

These are plane waves incident to plane interfaces, the vertical plane xz being the plane of symmetry. The problem is thus limited to two dimensions x and z. If Φ is the dilatational potential and $\overrightarrow{\Psi}$ the distortional potential, the displacement components u and w and the stress components σ_{zx} and σ_{zz} are written in each of the media:

$$\begin{cases} u = \dfrac{\partial \Phi}{\partial x} - \dfrac{\partial \Psi}{\partial z} \\[2mm] w = \dfrac{\partial \Phi}{\partial z} + \dfrac{\partial \Psi}{\partial x} \\[2mm] \sigma_{zx} = \mu\left(\dfrac{\partial w}{\partial x} + \dfrac{\partial u}{\partial z}\right) \\[2mm] \sigma_{zz} = \lambda\left(\dfrac{\partial u}{\partial x} + \dfrac{\partial w}{\partial z}\right) + 2\mu\,\dfrac{\partial w}{\partial z} \end{cases} \tag{2.53}$$

by simply denoting by Ψ the component Ψ_2 of the distortional potential.
The dilatational and distortional potentials are written:

(a) Incident dilatational wave: $\Phi_0 \exp\left(j(\omega t - lx + a_1 z)\right)$.
(b) Incident distortional wave: $\Psi_0 \exp\left(j(\omega t - lx + b_1 z)\right)$.
(c) Reflected dilatational wave: $\Phi_r \exp\left(j(\omega t - lx - a_1 z)\right)$.
(d) Reflected distortional wave: $\Psi_r \exp\left(j(\omega t - lx - b_1 z)\right)$.
(e) Transmitted dilatational wave: $\Phi_t \exp\left(j(\omega t - lx + a_2 z)\right)$.
(f) Transmitted distortional wave: $\Psi_t \exp\left(j(\omega t - lx + b_2 z)\right)$.

with

$$\begin{cases} l = \dfrac{\omega}{c} \\[3mm] c = \text{apparent velocity along the interface} \\[3mm] a_1 = \sqrt{\dfrac{\omega^2}{V_{p_1}^2} - l^2} \\[4mm] b_1 = \sqrt{\dfrac{\omega^2}{V_{s_1}^2} - l^2} \\[4mm] a_2 = \sqrt{\dfrac{\omega^2}{V_{p_2}^2} - l^2} \\[4mm] b_2 = \sqrt{\dfrac{\omega^2}{V_{s_2}^2} - l^2} \end{cases} \tag{2.54}$$

The potentials are expressed as follows:

(a) In the upper medium:

$$\left| \begin{aligned} \Phi_1 &= \Phi_0 \exp\left(j(\omega t - lx + a_1 z)\right) + \Phi_r \exp\left(j(\omega t - lx - a_1 z)\right) \\ \Psi_1 &= \Psi_0 \exp\left(j(\omega t - lx + b_1 z)\right) + \Psi_r \exp\left(j(\omega t - lx - b_1 z)\right) \end{aligned} \right. \tag{2.55}$$

(b) In the lower medium:

$$\left| \begin{aligned} \Phi_2 &= \Phi_t \exp\left(j(\omega t - lx + a_2 z)\right) \\ \Psi_2 &= \Psi_t \exp\left(j(\omega t - lx + b_2 z)\right) \end{aligned} \right. \tag{2.56}$$

with $\Psi_0 \equiv 0$ for an incident dilatational wave, and $\Phi_0 \equiv 0$ for an incident distortional wave.

The boundary conditions are obtained from Eqs. (2.53) by writing the continuity of the displacements and stresses on either side of the interface:

$$\left|\begin{array}{l} u_1 = u_2 \\ w_1 = w_2 \\ \sigma_{zx_1} = \sigma_{zx_2} \\ \sigma_{zz_1} = \sigma_{zz_2} \end{array}\right. \tag{2.57}$$

This gives the four Zoeppritz equations (Zoeppritz, 1919) which relate Φ_0, Ψ_0, Φ_r, Ψ_r, Φ_t and Ψ_t, and serve to compute the four reflection and transmission coefficients of the potentials.

- For an incident P-wave:
The P-wave potentials, reflection and transmission coefficients obtained are:

$$\frac{\Phi_r}{\Phi_0}, \quad \frac{\Phi_t}{\Phi_0}$$

and for the SV-wave potentials:

$$\frac{\Psi_r}{\Phi_0}, \quad \frac{\Psi_t}{\Phi_0}$$

- For an incident SV-wave:
The P-wave potentials, reflection and transmission coefficients obtained are:

$$\frac{\Phi_r}{\Psi_0}, \quad \frac{\Phi_t}{\Psi_0}$$

and for SV-wave potentials:

$$\frac{\Psi_r}{\Psi_0}, \quad \frac{\Psi_t}{\Psi_0}$$

These coefficients are functions of the incident angle θ_1, the densities ρ_1 and ρ_2, and the propagation velocities V_{p_1}, V_{s_1}, V_{p_2} and V_{s_2}. For incident angles greater than the critical angles, the apparent velocity c is less than V_{p_1}, V_{p_2} or V_{s_2}, and the quantities a_1, a_2 and b_2 become imaginary. The reflection and transmission coefficients are then complex, indicating the existence of phase shifts and evanescent waves.

Second case: incident SH-wave

The case of an incident SH-wave can be treated in the same way. It gives rise to a reflected SH-wave and a transmitted SH-wave, but no SH-wave to P-wave conversion occurs at the passage of the interface, in so far as the interface is parallel to the wave polarization direction.

2.7.2.2 Energy reflection and transmission coefficients

It can be shown (Ewing *et al.*, 1957, p. 81) that the energy reflection and transmission coefficients, the ratios of the energies reflected E_r and transmitted E_t to the incident energy

E_0, are expressed as a function of the reflection and transmission coefficients of the potentials by the formulas:

$$\left| \begin{aligned} \frac{E_r}{E_0} &= \left(\frac{\Phi_r}{\Phi_0}\right)^2 \frac{\tan \theta_0}{\tan \theta_r} \\ \frac{E_t}{E_0} &= \left(\frac{\Phi_t}{\Phi_0}\right)^2 \frac{\rho_2}{\rho_1} \frac{\tan \theta_0}{\tan \theta_t} \end{aligned} \right. \tag{2.58}$$

where

Φ	= dilatational or distortional potentials depending on each case,
θ_0, θ_r and θ_t	= incident, reflection and transmission angles,
ρ_1 and ρ_2	= densities of the two media.

Figure 2.6 shows, as a function of angle of incidence, the ratio of the reflected energy to the incident energy, for an incident P-wave and for different values of densities and propagation velocities in the two media. The reflection and transmission coefficients vary considerably according to the angle of incidence for incidences greater than 15 to 20°. On the other hand, they vary only slightly for low incidences (between 0 and 15°), making it possible, to a certain degree, to characterize the seismic interfaces by their reflectivity in reflection shooting.

2.7.2.3 Reflection and transmission coefficients in normal incidence

The reflection and transmission coefficients of plane waves are considerably simplified in normal incidence and yield familiar expressions:

First case: incident P-wave

Only the vertical component w of displacement and σ_{zz} of the normal stress subsist, with:

$$\left| \begin{aligned} w &= \frac{\partial \Phi}{\partial z} \\ \sigma_{zz} &= \rho V_p^2 \frac{\partial w}{\partial z} \end{aligned} \right. \tag{2.59}$$

By writing the boundary conditions at the interface:

$$\left| \begin{aligned} w_1 &= w_2 \\ \sigma_{zz_1} &= \sigma_{zz_2} \end{aligned} \right. \tag{2.60}$$

by taking the interface for the z origin and by deriving Eqs (2.55) and (2.56) with respect to z, we obtain:

$$\left| \begin{aligned} a_1(\Phi_0 - \Phi_r) &= a_2 \Phi_t \\ \rho_1 V_{p_1}^2 a_1^2 (\Phi_0 + \Phi_r) &= \rho_2 V_{p_2}^2 a_2^2 \Phi_t \end{aligned} \right. \tag{2.61}$$

Noting that:

$$a_1 = \frac{\omega}{V_{p_1}} \quad \text{and} \quad a_2 = \frac{\omega}{V_{p_2}}$$

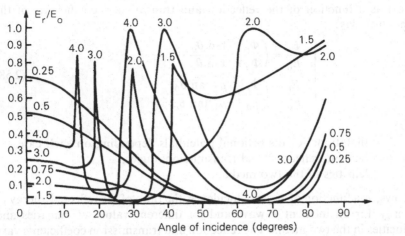

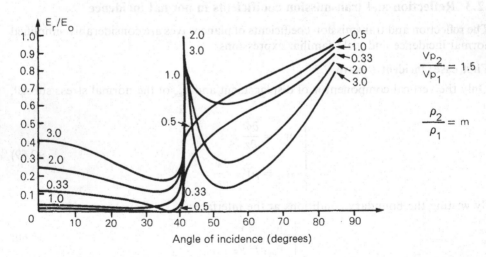

Fig. 2.6 Ratio of reflected energy to incident energy E_R/E_0 for a compressional wave incident upon the interface between two media 1 and 2, as a function of angle of incidence for different density ratios ρ_2/ρ_1 and velocity ratios V_{p_2}/V_{p_1}. Hypothesis of plane waves. Poisson's ratio $\sigma_1 = \sigma_2 = 0.25$ in both media. As the Poisson's ratio varies, E_R/E_0 may vary significantly. (From Sheriff, 1973, *Encyclopedic Dictionary of Exploration Geophysics*, Society of Exploration Geophysicists, Tulsa, Oklahoma, USA).

in normal incidence, we have:

$$\left|\begin{array}{l} \Phi_0 - \Phi_r = \Phi_t \dfrac{V_{p_1}}{V_{p_2}} \\[2mm] \Phi_0 + \Phi_r = \Phi_t \dfrac{\rho_2}{\rho_1} \end{array}\right. \tag{2.62}$$

giving the reflection and transmission coefficients of the potential in normal incidence:

$$\left|\begin{array}{l} \dfrac{\Phi_r}{\Phi_0} = \dfrac{\rho_2 V_{p_2} - \rho_1 V_{p_1}}{\rho_2 V_{p_2} + \rho_1 V_{p_1}} \\[3mm] \dfrac{\Phi_t}{\Phi_0} = \dfrac{2\rho_1 V_{p_2}}{\rho_2 V_{p_2} + \rho_1 V_{p_1}} \end{array}\right. \tag{2.63}$$

In practice, it is often preferable to use the reflection and transmission coefficients of the displacements or stresses. This easily yields the following:

(a) For displacements (vertical displacements):

$$\left|\begin{array}{l} \dfrac{w_r}{w_0} = \dfrac{\rho_1 V_{p_1} - \rho_2 V_{p_2}}{\rho_1 V_{p_1} + \rho_2 V_{p_2}} \\[3mm] \dfrac{w_t}{w_0} = \dfrac{2\rho_1 V_{p_1}}{\rho_1 V_{p_1} + \rho_2 V_{p_2}} \end{array}\right. \tag{2.64}$$

(b) For stresses (normal stresses):

$$\left|\begin{array}{l} \dfrac{\sigma_r}{\sigma_0} = \dfrac{\rho_2 V_{p_2} - \rho_1 V_{p_1}}{\rho_1 V_{p_1} + \rho_2 V_{p_2}} \\[3mm] \dfrac{\sigma_t}{\sigma_0} = \dfrac{2\rho_2 V_{p_2}}{\rho_1 V_{p_1} + \rho_2 V_{p_2}} \end{array}\right. \tag{2.65}$$

Second case: incident S-wave

In this case, only the horizontal components u of displacement and σ_{zx} of the tangential stress subsist with:

$$\left|\begin{array}{l} u = -\dfrac{\partial \Psi}{\partial z} \\[3mm] \sigma_{zx} = \rho V_s^2 \dfrac{\partial u}{\partial z} \end{array}\right. \tag{2.66}$$

By writing the boundary conditions at the interface:

$$\left|\begin{array}{l} u_1 = u_2 \\ \sigma_{zx_1} = \sigma_{zx_2} \end{array}\right. \tag{2.67} \\ \tag{2.68}$$

by taking the interface for the z origin and by deriving Eqs (2.55) and (2.56) with respect to z, we obtain:

$$\left|\begin{array}{l} b_1(\Psi_0 - \Psi_r) = b_2 \Psi_t \\ \rho_1 V_{s_1}^2 b_1^2 (\Psi_0 + \Psi_r) = \rho_2 V_{s_2}^2 b_2^2 \Psi_t \end{array}\right. \tag{2.69}$$

Noting that:

$$b_1 = \frac{\omega}{V_{s_1}} \quad \text{and} \quad b_2 = \frac{\omega}{V_{s_2}}$$

in normal incidence, we have:

$$\left|\begin{array}{l} \Psi_0 - \Psi_r = \Psi_t \dfrac{V_{s_1}}{V_{s_2}} \\[2ex] \Psi_0 + \Psi_r = \Psi_t \dfrac{\rho_2}{\rho_1} \end{array}\right. \tag{2.70}$$

which yield the reflection and transmission coefficients of the potential in normal incidence:

$$\left|\begin{array}{l} \dfrac{\Psi_r}{\Psi_0} = \dfrac{\rho_2 V_{s_2} - \rho_1 V_{s_1}}{\rho_2 V_{s_2} + \rho_1 V_{s_1}} \\[2ex] \dfrac{\Psi_t}{\Psi_0} = \dfrac{2\rho_1 V_{s_2}}{\rho_2 V_{s_2} + \rho_1 V_{s_1}} \end{array}\right. \tag{2.71}$$

The S-wave reflection and transmission coefficients obtained for the displacements and stresses are similar to those of P-waves by replacing V_p by V_s:

(a) For displacements (horizontal displacements):

$$\left|\begin{array}{l} \dfrac{u_r}{u_0} = \dfrac{\rho_1 V_{s_1} - \rho_2 V_{s_2}}{\rho_1 V_{s_1} + \rho_2 V_{s_2}} \\[2ex] \dfrac{u_t}{u_0} = \dfrac{2\rho_1 V_{s_1}}{\rho_1 V_{s_1} + \rho_2 V_{s_2}} \end{array}\right. \tag{2.72}$$

(b) For stresses (tangential stresses):

$$\left|\begin{array}{l} \dfrac{\sigma_r}{\sigma_0} = \dfrac{\rho_2 V_{s_2} - \rho_1 V_{s_1}}{\rho_1 V_{s_1} + \rho_2 V_{s_2}} \\[2ex] \dfrac{\sigma_t}{\sigma_0} = \dfrac{2\rho_2 V_{s_2}}{\rho_1 V_{s_1} + \rho_2 V_{s_2}} \end{array}\right. \tag{2.73}$$

2.8 SURFACE WAVES

In infinite, homogeneous and isotropic media, waves propagate in two possible ways: as compressional waves (velocity V_p), and as shear waves (velocity V_s). If the medium is not infinite, but bounded by a free surface (ground surface in seismic prospecting), other types of propagation may exist in the neighborhood of the surface, at velocities different from V_p and V_s. These relate to surface waves that propagate directly from the source to the detectors, without penetrating deeply into the subsurface, causing surface noise that may be disturbing and is liable to mask the reflections. We shall show below (Section 3.5) how appropriate transmission and reception systems help to eliminate them. In this Section, we shall merely analyze the nature of surface waves.

2.8.1 Semi-infinite, homogeneous and isotropic medium, Rayleigh waves

By definition, a surface wave propagates along the free surface and is rapidly attenuated with depth. In what conditions can a wave propagate in the x direction while attenuating in the z direction?

Let us assume that the problem has two dimensions x and z, and that the particle motion is contained in the vertical propagation plane (Fig. 2.7). We therefore have:

$$\left| \begin{array}{l} v \equiv 0 \\ u \text{ and } w \text{ independent of } y \end{array} \right.$$

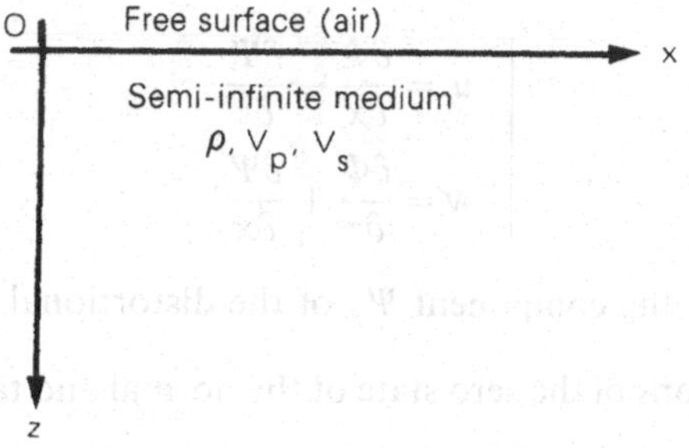

Fig. 2.7 Semi-infinite, homogeneous and isotropic medium. Two-dimensional problem x, z.

If Φ is the dilatational potential and $\vec{\Psi}$ the distortional potential, a solution can be written for a harmonic component of a surface wave with angular frequency ω, in the form:

$$\left| \begin{array}{l} \Phi = \Phi_0 \exp\left(j(\omega t - lx + az)\right) \\ \vec{\Psi} = \vec{\Psi}_0 \exp\left(j(\omega t - lx + bz)\right) \end{array} \right. \tag{2.74}$$

with

$$\left| \begin{array}{l} l = \dfrac{\omega}{c} \\[2mm] c = \text{propagation velocity along the surface} \\[2mm] a = j\sqrt{l^2 - \dfrac{\omega^2}{V_p^2}} \\[3mm] b = j\sqrt{l^2 - \dfrac{\omega^2}{V_s^2}} \end{array} \right. \tag{2.75}$$

and with the condition that a and b be pure imaginary, i.e.:

$$c < V_s < V_p$$

which states that the potentials tend towards zero as $z \to +\infty$ (z directed positively downward): this is the necessary condition for the wave to be attenuated with depth.

The solutions Φ and $\vec{\Psi}$ can be calculated by writing the boundary conditions on the free surface, which are expressed by the absence of surface stresses. For a problem with two dimensions, the zero state condition of the normal and tangential stresses is written:

$$\sigma_{zz} = 0 \quad \text{and} \quad \sigma_{zx} = 0 \quad \text{for} \quad z = 0$$

Let u and w be the components of particle motion at the surface. By definition we have:

$$\left|\begin{aligned} \sigma_{zz} &= \lambda \frac{\partial u}{\partial x} + (\lambda + 2\mu)\frac{\partial w}{\partial z} \\ \sigma_{zx} &= \mu\left(\frac{\partial u}{\partial z} + \frac{\partial w}{\partial x}\right) \end{aligned}\right. \tag{2.76}$$

with

$$\left|\begin{aligned} u &= \frac{\partial \Phi}{\partial x} - \frac{\partial \Psi}{\partial z} \\ w &= \frac{\partial \Phi}{\partial z} + \frac{\partial \Psi}{\partial x} \end{aligned}\right.$$

by simply denoting by Ψ the component Ψ_2 of the distortional potential.

This gives the expressions of the zero state of the normal and tangential stresses at the surface:

$$\left|\begin{aligned} \sigma_{zz} &= \lambda(l^2\Phi_0 + lb\Psi_0) + (\lambda + 2\mu)(a^2\Phi_0 - lb\Psi_0) = 0 \\ \sigma_{zx} &= \mu[2la\Phi_0 + (b^2 - l^2)\Psi_0] = 0 \end{aligned}\right. \tag{2.77}$$

By replacing $\lambda + 2\mu$ by ρV_p^2, μ by ρV_s^2, and a, b and l by their values as a function of c, V_p and V_s, the boundary conditions obtained are:

$$\left|\begin{aligned} \Phi_0\left(2 - \frac{C^2}{V_s^2}\right) + 2j\Psi_0\sqrt{1 - \frac{C^2}{V_s^2}} &= 0 \\ 2j\Phi_0\sqrt{1 - \frac{C^2}{V_p^2}} - \Psi_0\left(2 - \frac{C^2}{V_s^2}\right) &= 0 \end{aligned}\right. \tag{2.78}$$

To obtain non-zero solutions for Φ_0 and Ψ_0, the matrix of the system must be singular and the following condition is obtained:

$$\left|\left(2 - \frac{C^2}{V_s^2}\right)^2 = 4\sqrt{1 - \frac{C^2}{V_p^2}}\sqrt{1 - \frac{C^2}{V_s^2}}\right. \tag{2.79}$$

This is the relationship between the velocity C of a surface wave in a semi-infinite and homogeneous medium, the dilatational velocity V_p and the distortional velocity V_s.

Condition (2.79) is called the Rayleigh equation, after the famous English physicist of the late 19th century, who was the first to formulate it. There is only one real solution C_R satisfying the condition $0 < C < V_s < V_p$. This surface wave propagating in the neighborhood of the free surface at velocity C_R is called the **Rayleigh wave**.

The harmonic component with angular frequency ω is defined by:

$$\Phi = \Phi_0 \exp\left(j\omega\left(t - \frac{x}{C_R}\right)\right) \exp\left(-\sqrt{l^2 - \frac{\omega^2}{V_p^2}}\, z\right)$$

$$\Psi = \Psi_0 \exp\left(j\omega\left(t - \frac{x}{C_R}\right)\right) \exp\left(-\sqrt{l^2 - \frac{\omega^2}{V_s^2}}\, z\right)$$

$$(2.80)$$

where Φ_0 and Ψ_0 are the solutions of Eqs (2.78).

In semi-infinite homogeneous media, the Rayleigh wave propagation velocity C_R does not depend on the frequency, but only on the propagation velocities V_p and V_s, in other words on the moduli of elasticity of the medium. It can be shown (Ewing, Jardetzky and Press, 1957, p. 34) that the ratio C_R/V_s increases more or less linearly with the Poisson's ratio σ:

$$
\begin{aligned}
\frac{C_R}{V_s} &= 0.875 && \text{for } \sigma = 0 \\
&= 0.919 && \text{for } \sigma = 1/4 \\
&= 0.955 && \text{for } \sigma = 1/2
\end{aligned}
\qquad (2.81)
$$

The dilatational and distortional potentials of Rayleigh waves decrease exponentially with depth, and the particle motion decreases with depth as the sum of two exponentials. The particle motion is contained in the vertical propagation plane. In the neighborhood of the surface, it is elliptical and retrograde in plane xz, with a vertical displacement about 1.5 times the horizontal displacement. Horizontal motion is nullified at a depth of about 0.2 times the wavelength c/f, and becomes elliptical and prograde at greater depth.

In the two-dimensional case, the Rayleigh wave propagates without geometric spreading along the free surface. In practice, the problem is three-dimensional: the Rayleigh waves are annular about the point source, and it can be shown that they are attenuated with distance r according to a law close to $r^{-1/2}$.

As a rule, Rayleigh waves are generated whenever a P- or SV-wave is generated in the neighborhood of a free surface. Particle motion lies in the vertical propagation plane, and decreases with depth as the sum of two exponentials. Rayleigh waves propagate along the free surface at a velocity slightly lower than the shear wave velocity.

2.8.2 Semi-infinite, homogeneous and isotropic medium overlaid by a surface layer of thickness h. Dispersion

This is a more realistic case than that of the semi-infinite homogeneous medium, because, in nature, a surface layer often exists, weathered to varying degrees, from a few meters to some 20 m thick, that can play the role of a waveguide for surface waves. This is the Weathered Zone (WZ).

2.8.2.1 Pseudo-Rayleigh waves

In the presence of a surface layer, waves similar to the Rayleigh waves can propagate along the surface and cause considerable surface noise that is often called "ground roll".

These are pseudo-Rayleigh waves or dispersive Rayleigh waves. While Rayleigh waves in semi-infinite homogeneous media have a propagation velocity that is independent of the frequency, pseudo-Rayleigh waves generally exhibit a dispersive character.

Dispersion

Dispersion is the deformation of a wavetrain due to the variation in propagation velocity with frequency. For pseudo-Rayleigh waves, it can be shown that the high frequencies, that correspond to short wavelengths in comparison with thickness h, propagate at the velocity C_{R_1} of Rayleigh waves in the upper layer, and the low frequencies, with long wavelengths in comparison with h, at the velocity C_{R_2} of Rayleigh waves in the lower medium (Fig. 2.8).

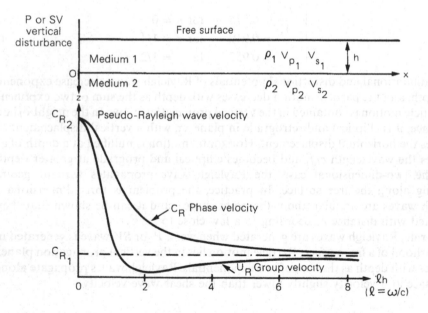

Fig. 2.8 Semi-infinite, homogeneous and isotropic medium overlaid by a layer of thickness h. Dispersion curves of pseudo-Rayleigh waves. Two-dimensional problem, x, z.

Dispersion is accompanied by a separation of the phase velocity and the group velocity. The **phase velocity** is the distance traveled per unit time by a constant-phase point of the wave surface (for example a maximum or a minimum). The **group velocity** is the propagation velocity of the wavetrain envelope. In the absence of dispersion, the phase and group velocities are equal. In the presence of dispersion, it can be shown (Ewing *et al.*, p. 68) that they are related by the equation:

$$U = C + l \frac{dC}{dl} \tag{2.82}$$

where

$$C = \frac{\omega}{l} \text{ phase velocity,}$$

$$U = \frac{d\omega}{dl} \text{ group velocity,}$$

$l = 2\pi/\text{wavelength.}$

The dispersion of Rayleigh waves generally causes a change in the shape of the wavetrain with distance.

Example: Figure 2.9 shows a recording of dispersive surface waves. In seismic recordings, the signals received by the different detectors are generally juxtaposed, in order to form a raster of traces whose origins are positioned like the detectors along the surface. In the case at hand, the source is located near the right-hand detector, and the geophones are placed along line Ox between 0 and 450 m at a spacing of one geophone every 5 m. By definition, line Ox, which connects the source to the geophones, is the "profile". The arrival time of the seismic signals is shown vertically with a scale of 10 cm/s.

In this recording, a very high surface noise can be distinguished, which crosses the section obliquely and is characterized by very low frequencies (6 to 12 Hz). The wavetrain starts from the neighborhood of the source (right-hand part of the recording), where the disturbance time ranges between 0 and 0.4 s, and propagates as far as the furthest geophones (left-hand part of the recording) where the disturbance begins from 1 s and continues after 2 s.

This surface noise mainly consists of Rayleigh waves and dispersive pseudo-Rayleigh waves. Figure 2.9 shows two families of surface waves: a first family between 0 and 100 m from the source, whose phase and group velocities are relatively slow (200 to 300 m/s), and a second family between 100 and 400 m, whose velocities are faster (500 to 600 m/s). The dispersive character of the first family is clearly visible, with a group velocity significantly lower than the phase velocity.

The seismic arrivals preceding the surface waves (between 0 and 1 s at the top left-hand of the recording) are reflections and refractions. The arrival at higher frequency (70 Hz), which crosses the recording obliquely between 0 and 1.35 s, is an air wave, with a propagation velocity of 340 m/s.

2.8.2.2 Love waves

Love waves, named after an English engineer of the early 20th century, are SH-type waves, namely polarized perpendicular to the vertical plane of the source/geophone system, which propagate along the surface and are dispersive in the presence of a surface layer. The equation of Love waves is obtained by writing the boundary conditions for the propagation of an SH-wave at the surface of a semi-infinite medium overlaid by a surface layer of thickness h (Fig. 2.10).

Let us take the base of the surface layer for the Ox axis, and select the Oz axis directed downward. Let us consider the problem with two dimensions x and z, and assume that the excitation source is in the neighborhood of the surface in medium 1 and directed along Oy.

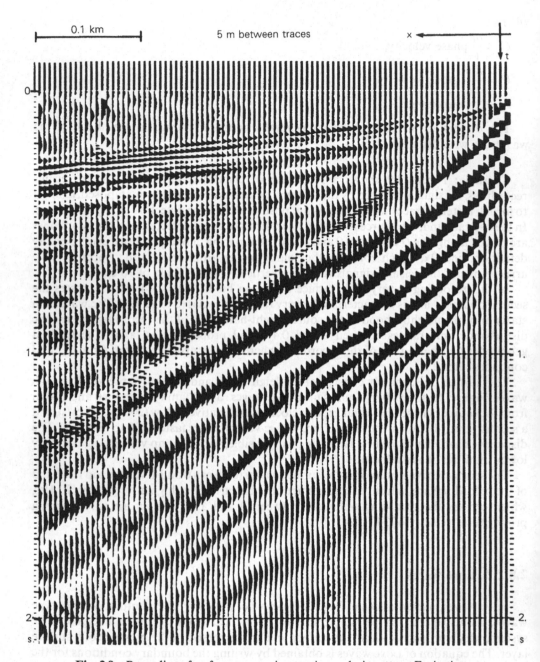

0.1 km   5 m between traces   x

Fig. 2.9 Recording of surface waves using a noise analysis pattern. Excitation by vertical disturbance at the right of the figure : dropping of a 400 kg weight on a target coupled to the ground. One geophone per trace. Ninety geophones positioned over 450 m. 5 m between geophones. Recording time 2 s. On the recorded signals ("traces"), the positive parts have been blacked in and the negative parts removed. This is the "variable area" display. (*IFP* Document).

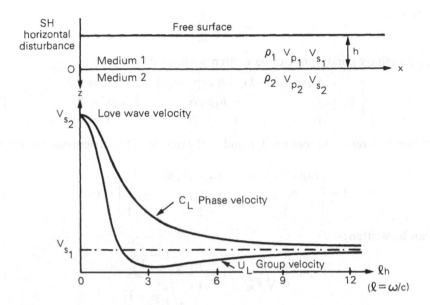

Fig. 2.10 Semi-infinite, homogeneous and isotropic medium overlaid by a layer of thickness h. Dispersion curves of Love waves. Two-dimensional problem x, z.

Particle motions v_1 and v_2 in media 1 and 2 hence occur along Oy, and can be written:

$$
\begin{vmatrix}
v_1 = A \exp\left(j(\omega t - lx + b_1 z)\right) + B \exp\left(j(\omega t - lx - b_1 z)\right) \\
v_2 = C \exp\left(j(\omega t - lx + b_2 z)\right)
\end{vmatrix}
\tag{2.83}
$$

with

$$
\begin{vmatrix}
l &= \dfrac{\omega}{c} \\[2mm]
b_1 &= \sqrt{\dfrac{\omega^2}{V_{s_1}^2} - l^2} = l\sqrt{\dfrac{c^2}{V_{s_1}^2} - 1} \\[2mm]
b_2 &= \sqrt{\dfrac{\omega^2}{V_{s_2}^2} - l^2} = jl\sqrt{1 - \dfrac{c^2}{V_{s_2}^2}}
\end{vmatrix}
\tag{2.84}
$$

and with the condition

$$
V_{s_1} < c < V_{s_2}
\tag{2.85}
$$

This condition must be satisfied for the wave to be attenuated exponentially with depth in medium 2.

The boundary conditions are written as follows:

(a) Zero stresses at the free surface: $(\sigma_{zy})_1 = 0$ for $z = -h$.
(b) Continuity of stresses at the interface: $(\sigma_{zy})_1 = (\sigma_{zy})_2$ for $z = 0$.
(c) Continuity of displacements at the interface $v_1 = v_2$ for $z = 0$.

Noting that :

$$\sigma_{zy} = \mu \frac{\partial v}{\partial z}$$

the three boundary conditions can be written without difficulty:

$$\left\{ \begin{array}{lll} A \exp(-jb_1 h) - B \exp(jb_1 h) & = 0 \\ b_1 \mu_1 A \qquad\qquad - b_1 \mu_1 B & + b_2 \mu_2 C = 0 \\ A \qquad\qquad + \qquad B & - \qquad C = 0 \end{array} \right. \qquad (2.86)$$

To obtain non-zero solutions for A, B and C, the matrix of the system must be singular, namely:

$$\Delta = \begin{vmatrix} \exp(-jb_1 h) & -\exp(jb_1 h) & 0 \\ b_1 \mu_1 & -b_1 \mu_1 & b_2 \mu_2 \\ 1 & 1 & -1 \end{vmatrix} = 0 \qquad (2.87)$$

which can be written:

$$\tan \frac{\omega h}{c} \sqrt{\frac{c^2}{V_{s_1}^2} - 1} = \frac{\mu_2}{\mu_1} \frac{\sqrt{1 - \frac{c^2}{V_{s_2}^2}}}{\sqrt{\frac{c^2}{V_{s_1}^2} - 1}} \qquad (2.88)$$

where

c	= Love wave propagation velocity,
h	= thickness of surface layer,
μ_1 and μ_2	= rigidity moduli in the surface layer and in the lower medium,
V_{s_1} and V_{s_2}	= shear velocities.

Equation (2.88) is the Love equation (1911). It allows an infinity of solutions C_L which correspond to the different propagation modes of the SH-wave channelled by layer h. It can be shown that real square roots only exist for:

$$V_{s_1} < c < V_{s_2} \qquad (2.89)$$

Equation (2.88) shows that the solutions C_L depend on the frequency $f = \omega/2\pi$ and on the thickness h. Love waves thus display a dispersive character, like pseudo-Rayleigh waves. Their propagation velocity C_L varies with the wavelength. The high frequencies propagate at the shear wave velocity V_{s_1} in the upper medium, the low frequencies at the shear wave velocity V_{s_2} in the lower medium. As for the pseudo-Rayleigh waves, the dispersion of Love waves is accompanied by a separation between the phase velocity and the group velocity (Fig. 2.10). Pseudo-Rayleigh waves and Love waves often cause very high surface noise in seismic surveys. They must be eliminated by elaborate transmission and reception systems that are discussed below (Section 3.5).

2.9 DIFFRACTION

The laws of reflection and transmission stated in Section 2.7 are valid for plane and large-sized interfaces. They are generally invalid in the presence of faults, pinchouts, interrupted reflectors, and, as a rule, for all subsurface features that are small in

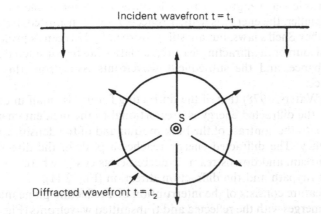

Fig. 2.11 Diffraction by a small-diameter sphere S. The incident wavefront is shown at time t_1 and the diffracted wavefront at time $t_2 > t_1$.

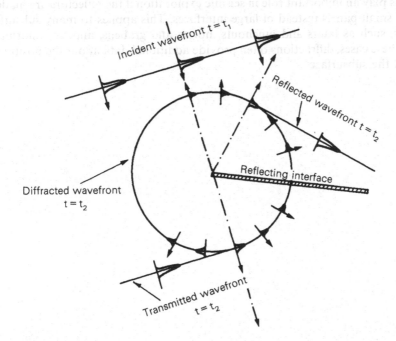

Fig. 2.12 Diffraction by an interrupted plane reflecting interface. Problem with two dimensions. The incident wavefront is shown at time t_1, and the transmitted, reflected and diffracted wavefronts at time $t_2 > t_1$. The amplitude of particle motion at the passage of the wavefronts is shown schematically.

comparison with seismic wavelengths. If these features are comparable in size to the wavelengths or are smaller, the energy is neither reflected nor transmitted, but diffracted. Diffractions do not obey Snell's laws, but are still governed by Huygens's principle: if the incident wavefront encounters a diffracting feature, the latter can be considered as a source of secondary disturbance, and the subsequent wavefronts issue from this source of secondary disturbance.

It can be shown (Waters, 1978) that, if the diffracting feature is small in comparison with the wavelength, the diffracted energy is proportional to the incident energy, to the volume of the feature, to the contrast of the bulk moduli and of the densities, and to the square of the frequency. The diffracted energy reaches a peak in the direction of the incident raypath upstream and downstream. It decreases as $\cos \beta$, where β is the angle between the incident raypath and the diffraction direction (Fig. 2.11).

If the diffracting feature consists of the interruption of a reflecting plane interface, the diffracted wavefront merges with the reflected and transmitted wavefronts (Fig. 2.12). The diffracted wavefront intensities are symmetrical about the incident raypath and the reflected raypath, but the amplitudes have opposite phases. **The diffracted and transmitted wavefronts merge** with continuity, and the amplitude of the diffracted wave at the boundary of the brightness zone and the shadow zone is equal to that of the direct wave decreased by that of the diffracted wave on the brightness zone. Similarly, **the diffracted and reflected wavefronts** merge with continuity.

Diffractions play an important role in seismic exploration if the reflectors are made of successions of small panels instead of large interfaces. This applies to many subsurface configurations, such as faults and pinchouts, horsts and grabens, and unconformable interfaces. In these cases, diffractions often provide additional data about the position of the features in the subsurface.

CHAPTER

3

seismic signals

3.1 SIGNAL AND NOISE. DEFINITIONS

It is not easy to give a definition of the seismic signal, because the separation of what is called the signal from the noise is strongly dependent on the problems analyzed. As a rule, the term "seismic signal" is applied to all recorded events from which it is hoped to derive information about the subsurface structure and geology. They include the succession of echos in reflection surveys, arrivals of conical waves in refraction surveys, and even certain diffractions which may facilitate the location of subsurface features. Anything not actually considered as signal is considered as noise. The following distinctions are drawn for noise:

(a) **Ambient and industrial noise,** which exists in recordings, even in the absence of seismic emission. Ambient noises are often disorganized noises, of a more or less incoherent nature (wind, microseisms). Industrial noises are generally more organized, such as the seismic noise due to automobile traffic, or the induction at 50 Hz (60 Hz in the USA) generated in conventional seismic cables in the neighborhood of high-voltage lines.

(b) **Source noise,** which is generated by the seismic emission. In onshore seismic prospecting, surface waves are often considered as source noise, although they may in certain cases contain useful data for interpretation. In offshore seismic prospecting, the diffractions generated when the incident wave encounters the unevennesses of the seafloor are often very disturbing noises. Diffractions caused by subsurface discontinuities, even if they can help to identify the position of certain features, are often considered as noise, because they occur in the form of hyperbolas in the time/distance representation, which partially mask the useful reflections. Also sometimes considered as noise are refracted arrivals in reflection surveys, converted waves ($P \to SV$), air waves, direct waves propagating in the water, etc.

3.2 SEISMIC PULSE

The seismic pulse is the elementary wavetrain emitted by the disturbance source. The **disturbance** is often created onshore by an explosion, a weight drop, or a vibrator, and offshore by the release of a volume of air or water into the water, the variation in volume of a submerged body, an implosion, a spark, etc. The duration of the elementary wavetrain is relatively short, except in vibroseismic prospecting. It is normally in the range of a few tens of milliseconds, to allow separation of the echos from the successive interfaces with sufficient resolution. The seismic pulse, which is sometimes called a wavelet, generally contains one to three oscillation cycles (Figs 3.1 and 3.2).

In vibroseismic prospecting, the emission uses vibrators and the disturbance may last several seconds. The emitted wavetrain is a continuously varied frequency signal ("sweep") and, as we shall show below (Section 3.6), it must be given specific properties in order to obtain, after processing, a resolution level equivalent to that of the short pulses. A convenient (but not always feasible) way to identify the seismic pulse is to record a reflection from a single reflector, sufficiently far from the other reflectors to avoid interference with the other reflections and with surface waves.

Figure 3.1 shows two reflected seismic pulses, the first obtained onshore by a weight drop, and the second offshore by an imploder. In both cases, the pulse corresponds to a signal reflected from a single reflector located at a depth of about 1 km. The signal has been detected respectively by a geophone which delivers an electric voltage proportional to the ground particle velocity (about 50 mV/mm s^{-1}) and by a hydrophone, which delivers an electric voltage proportional to the variation in pressure in the water (about 1 mV/mbar). The seismic pulse can also be measured in the neighborhood of the source, before propagation in the subsurface. To do this, a geophone and a hydrophone are placed immediately next to the source (a few meters), giving the "signature" of the source independent of the subsurface as a first approximation (Fig. 3.2).

The relative deformation of the rock due to the seismic wave is low, usually less than 10^{-8}, except in the neighborhood of the disturbance source. For a wavelength of 50 m, for example, it corresponds to a particle displacement of 0.125 μm.

3.3 DEFINITIONS OF FOURIER TRANSFORM, CONVOLUTION, CORRELATION AND AUTOCORRELATION

We have seen that the seismic signal is a succession of events recorded as a function of time. Another way of representing the data contained in the seismic signal exists, the representation in the "harmonic" domain, as a function of frequency. The passage from the time domain to the harmonic domain is carried out by the **Fourier transform**.

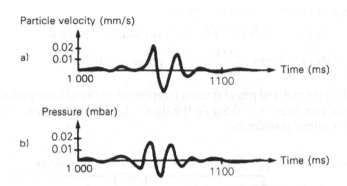

Fig. 3.1 Seismic pulses recorded by reflection upon a 1 km deep reflector.
(a) Onshore, and (b) Offshore.

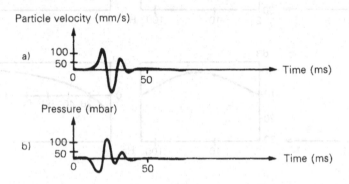

Fig. 3.2 Seismic pulses recorded a few meters from the source.
(a) Onshore, and (b) Offshore.

a. Fourier transform

It can be recalled that, if $s(t)$ is a signal, its Fourier transform $S(f)$ is written by definition:

$$S(f) = \int_{-\infty}^{+\infty} s(t) \exp(-2\pi j f t)\, dt \qquad (3.1)$$

where $S(f)$ is the representation of the signal in the harmonic domain.

The return to the time domain takes place by the inverse Fourier transform:

$$s(t) = \int_{-\infty}^{+\infty} S(f) \exp(2\pi j f t)\, df \qquad (3.2)$$

$S(f)$ is called the frequency spectrum of the signal $s(t)$. It is generally a complex function, containing a real part and an imaginary part:

$$S(f) = R(f) + jI(f)$$

The modulus and phase of $S(f)$ are written:

$$|S(f)| = \sqrt{R^2(f) + I^2(f)} \qquad \text{amplitude spectrum} \qquad (3.3)$$

$$\Phi(f) = \text{arc tan } \frac{I(f)}{R(f)} \qquad \text{phase spectrum} \qquad (3.4)$$

The amplitude spectrum and phase spectrum represent the amplitude and phase of the different sinusoïdal components making up the signal $s(t)$. Figure 3.3 shows a number of signals and their Fourier transforms.

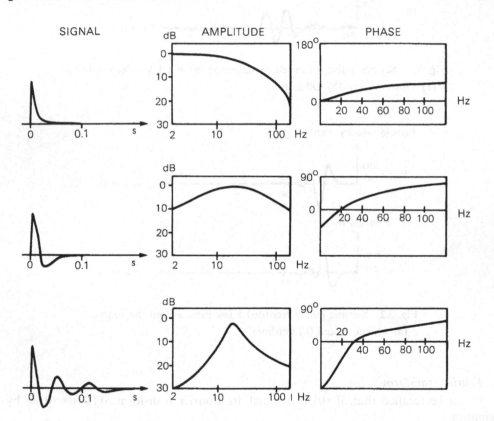

Fig. 3.3 Frequency spectra of three different signals. (From Kramer *et al.*, 1968, *Seismic Energy Sources Handbook*, Bendix United Geophysical, Pasadena, California, USA).

The Fourier transform plays a vital role in seismic prospecting, as in many other scientific and technical fields. The breakdown of a seismic recording into its different sinusoidal components makes it possible to carry out frequency filterings designed to attenuate the spurious noise and to strengthen the signal. It is possible, for example, to compute the Fourier transform of the noisy signal, filter by attenuating the sinusoidal components of the noise, and take the inverse Fourier transform which restores a less noisy signal.

b. Convolution

Filtering can also be carried out directly in the time domain, by the convolution operation. By definition, the convolution function of $g(t)$ by $s(t)$ is written:

$$c(t) = \int_{-\infty}^{+\infty} s(\tau)g(t - \tau) \, d\tau \tag{3.5}$$

It is often represented by the symbolic expression:

$$c(t) = s(t) * g(t) \tag{3.6}$$

The convolution of a signal $s(t)$ by a function $g(t)$ amounts to passing $s(t)$ through a linear filter whose impulse response is $g(t)$. The Fourier transform $G(f)$ of the impulse response of a filter is the transfer function of the filter. The impulse response hence characterizes the filter in the time domain, and the transfer function in the frequency domain.

It can be shown that the Fourier transform of a convolution of two functions is the product of the Fourier transforms of the two functions:

$$C(f) = S(f)G(f) \tag{3.7}$$

The amplitude spectrum is the product of the amplitude spectra, and the phase spectrum is the sum of the phase spectra of the two functions. In **digital recording**, the convolution of two numerical functions s_i and g_i discretized with a sampling interval $\Delta\tau$ is written:

$$c_j = \Delta\tau \sum_{i=0}^{n} s_i g_{j-i} \tag{3.8}$$

c. Cross-correlation

By definition, the cross-correlation function of $y(t)$ by $z(t)$ is written:

$$r(t) = \int_{-\infty}^{+\infty} y(\tau)z(t + \tau) \, d\tau \tag{3.9}$$

which is symbolically written:

$$r(t) = y(t) * z(-t) \tag{3.10}$$

The cross-correlation function is characteristic of the degree of similarity of the functions $y(t)$ and $z(t)$: if $r(t)$ is relatively large, the similarity is good; if $r(t)$ is close to zero the similarity is poor. The cross-correlation function is often standardized by the denominator:

$$D = \left[\int_{-\infty}^{+\infty} y^2(\tau) \, d\tau \int_{-\infty}^{+\infty} z^2(\tau) \, d\tau \right]^{1/2} \tag{3.11}$$

and $r(0)/D = 1$ is obtained for the perfect identity of the functions y and z.

It can be shown that the Fourier transform of the cross-correlation of two functions is the product of the Fourier transform of the first, multiplied by the conjugate Fourier transform of the second:

$$R(f) = Y(f)\overline{Z}(f) \tag{3.12}$$

The amplitude spectrum is the product of the amplitude spectra, and the phase spectrum is the difference of the phase spectra of the two functions. In **digital recording,** the correlation of two numerical functions y_i and z_i discretized with a sampling interval $\Delta\tau$ is written:

$$r_j = \Delta\tau \sum_{i=0}^{n} y_i z_{j+i} \tag{3.13}$$

d. Autocorrelation

By definition, the autocorrelation function of $s(t)$ is written:

$$a(t) = \int_{-\infty}^{+\infty} s(\tau)s(t + \tau)\, d\tau \tag{3.14}$$

or symbolically:

$$a(t) = s(t) * s(-t) \tag{3.15}$$

Its Fourier transform is:

$$A(f) = S(f)\overline{S}(f) \tag{3.16}$$

Its amplitude spectrum is the square of the amplitude spectrum of $s(t)$. Its phase spectrum is zero.

In **digital recording,** the autocorrelation of a numerical function s_i discretized with a sampling interval $\Delta\tau$ is written:

$$a_j = \Delta\tau \sum_{i=0}^{n} s_i s_{j+i} \tag{3.17}$$

We shall show below (Section 3.6.2) that the cross-correlation and autocorrelation functions play a vital role in signal compression in vibroseismic prospecting.

3.4 ANALYTIC SIGNAL

The concept of analytic signal, frequently employed in telecommunications in signal transmission problems in amplitude and frequency modulation, is often used by seismic specialists (Chapel, 1980). Let $s(t)$ be a causal seismic signal ($h(t) = 0$ for $t \leqslant 0$), and $g(t)$ the quadrature signal (same amplitude spectrum as $s(t)$, and phase spectrum shifted by 90°). The analytic signal of $s(t)$ is by definition the complex function:

$$A(t) = s(t) + jg(t) \tag{3.18}$$

where $s(t)$ and $g(t)$ form a pair of Hilbert transforms.

We can write:

$$A(t) = M(t) \exp(j\Phi(t)) \tag{3.19}$$

with

$$M(t) = \sqrt{s^2(t) + g^2(t)} \tag{3.20}$$

$$\Phi(t) = \text{arc tan} \frac{g(t)}{s(t)} \tag{3.21}$$

where

$M(t)$ = **modulus** or **instantaneous amplitude** of the analytic signal,

$\Phi(t)$ = **instantaneous phase.**

The instantaneous angular frequency is: $\omega = \dfrac{\mathrm{d}\Phi}{\mathrm{d}t}$.

The **instantaneous frequency** is: $f(t) = \dfrac{1}{2\pi}\dfrac{\mathrm{d}\Phi}{\mathrm{d}t}$.

Relatively simple processes are available for computing the quadrature signal $g(t)$ corresponding to a signal $s(t)$ (Ville, 1948). The concept of analytic signal can be useful for determining the instantaneous amplitude in seismic recordings, which sometimes indicates the presence of hydrocarbons, the instantaneous phase which accurately indicates the continuity of the seismic horizons, and the instantaneous frequency, which serves in certain cases to estimate the absorption coefficient in the formations.

3.5 SURFACE NOISE FILTERING

We have shown that surface noises are often generated by the seismic source and propagate along the ground surface without penetrating deeply into the subsurface. They are repeated at each disturbance, but contain no information about the deep structures, and the high amplitudes of their oscillations, which may last several seconds, often mask the useful signals. These noises can be attenuated by geophone patterns aligned along the source/geophone direction or spread out laterally, and grouped on the same seismic channel. As a rule, surface noise can be attenuated by the combination of source arrays and geophone patterns aligned with the profile direction or spread out on the surface.

The geometry of the patterns must be selected in accordance with the characteristics of the signal and the noise, and particularly of their wavenumber spectra.

3.5.1 Wavenumber spectrum

In the same way as the frequency spectrum $S(f)$ is the Fourier transform of the seismic signal $s(t)$ in time, the wavenumber spectrum $S(k)$ is the Fourier transform of the seismic signal $s(x)$ in space. By definition:

$$S(k) = \int_{-\infty}^{+\infty} s(x) \exp\left(-2\pi jkx\right) \mathrm{d}x \tag{3.22}$$

The return to the space domain is carried out by the inverse Fourier transform:

$$s(x) = \int_{-\infty}^{+\infty} S(k) \exp\left(2\pi jkx\right) \mathrm{d}k \tag{3.23}$$

If c is the apparent propagation velocity in direction x and f is the frequency, the wavenumber here is expressed as a function of c and f by the equation:

$$k = \frac{f}{c} = \frac{l}{2\pi}$$
(3.24)

In practice, the wavenumber is expressed in cycles per meter or per kilometer.

3.5.2 Wavenumber filter

Geophone patterns introduce wavenumber filterings, attenuating certain wavenumbers and strengthening others. For example, the waves ascending vertically from the subsurface simultaneously reach all the geophones and are strengthened, while the waves propagating along the surface give rise to interferences that are destructive to varying degrees. For waves ascending from the reflectors, the apparent velocity along the surface is practically infinite, and the wavenumbers approach zero. For waves propagating along the surface, the wavenumbers are higher, around 20 to 40 cycles/km.

To strengthen the reflections and attenuate surface noise, the wavenumber filter should therefore strengthen the wavenumbers close to zero, and attenuate the high wavenumbers, greater than 20 cycles/km, for example. We shall show that this can be achieved by grouping a pattern of geophones on each seismic channel, aligned in the profile direction.

a. Transfer function of wavenumber filters

Consider n equidistant geophones aligned on the surface along the profile direction. If a plane wavefront reaches the surface with an incidence angle β and if $A_0 \exp{(j\omega t)}$ is the harmonic component of angular frequency ω recorded at the first geophone (Fig. 3.4), the successive geophones will record a signal:

$$A_i = A_0 \exp{(j(\omega t - lid))}$$
(3.25)

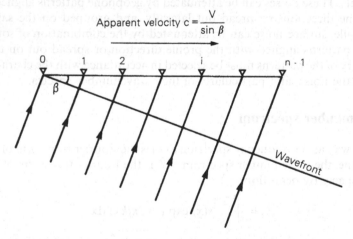

Fig. 3.4 Principle for computing wavenumber filtering obtained by a pattern of n equidistant geophones aligned with the profile. The vertical plane of the profile is the plane of symmetry. The signals of the n geophones are summed up in recording.

where

$l = \omega/c$,

c = apparent horizontal velocity,

i = geophone number (the first one has the number zero),

d = geophone spacing.

The signal obtained by grouping n geophones on a single channel is written:

$$\sum_{i=0}^{n-1} A_i = A_0 \exp(j\omega t) \sum_{i=0}^{n-1} \exp(-jlid)$$

$$= A_0 \exp(j\omega t) \frac{1 - \exp(jlnd)}{1 - \exp(jld)} \tag{3.26}$$

The transfer function of the wavenumber filter obtained by grouping the n geophones is hence:

$$H(k) = \frac{1 - \exp(jlnd)}{1 - \exp(jld)} \tag{3.27}$$

It can be written as a function of the wavenumber $k = l/2\pi$:

$$H(k) = \exp(j\pi k(n-1)d) \frac{\sin n\pi dk}{\sin \pi dk} \tag{3.28}$$

The modulus $|H(k)|$ is shown in Fig. 3.5. The first zero crossing occurs at $k = 1/nd$ and the period is $1/d$.

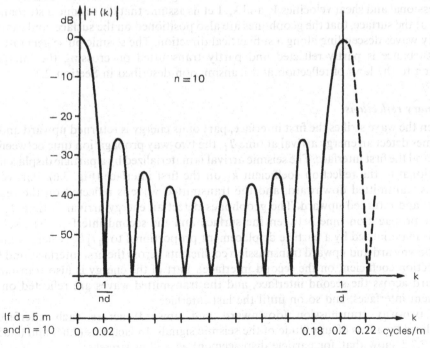

Fig. 3.5 Modulus $|H(k)|$ of the transfer function of the wavenumber filter, for $n = 10$ geophones.

b. *Example of a surface noise filtering pattern*

If, for example, $n = 10$ geophones are positioned at intervals of $d = 5$ m, $nd = 50$ m, the first zero crossing occurs at $k = 0.02$ cycle/m, and the periodicity is 0.2 cycle/m. The waves whose wavenumber is less than 0.02 cycle/m are only slightly attenuated, whereas those whose wavenumber is between 0.02 and 0.18 cycle/m are weakened. The wavenumbers between 0.18 and 0.22 cycle/m are again only slightly attenuated due to the periodicity of $H(k)$. For frequencies of 20 Hz, for example, waves whose apparent velocity along the surface is greater than 1000 m/s $(k < 20/1000)$ are only slightly attenuated (this is the case for reflected waves), and those whose apparent velocity ranges between 111 and 1000 m/s $(20/1000 < k < 20/111)$ are weakened. Those whose apparent velocity lies between 91 and 111 m/s $(20/111 < k < 20/91)$ are but slightly attenuated.

If the frequency of the surface waves is around 20 Hz, and their velocity ranges between 200 and 1000 m/s, the pattern selected will succeed in filtering the surface waves, without destroying the reflections from the subsurface.

3.6 REFLECTION SEISMOGRAM

It is possible to predict the succession of echos generated by reflection shooting for a layered half space. Let us consider a succession of geological beds with density ρ and compressional and shear velocities V_p and V_s. Let us assume that the seismic disturbance is created at the surface, that the geophones are also positioned on the surface, and examine the body waves descending along a subvertical direction. The seismic pulse generated by the disturbance is partly reflected and partly transmitted on crossing the interfaces, according to the laws of reflection and transmission described in Section 2.7.

a. *Primary reflections*

When the wave strikes the first interface, part of its energy is returned upward and the geophones detect an energy arrival at time T_1, the two-way propagation time between the surface and the first interface. The seismic arrival is materialized by a particle displacement proportional to the reflection coefficient k_1 on the first interface (Fig. 3.6). Part of the energy is transmitted downward, and the transmitted wave is reflected on the second interface and returned upward. The geophones detect an energy arrival at time T_2, the two-way propagation time between the surface and the second interface. The seismic arrival is materialized by a particle displacement proportional to $t_1 t_1' k_2$ (where t_1 and t_1' are the downward and upward transmission coefficients across the first interface, and k_2 is the reflection coefficient on the second interface). Part of the energy is also transmitted downward across the second interface, and the transmitted waves are reflected on the subsequent interfaces, and so on until the last interface.

The two-way transmission (downward and upward) across each interface is materialized by a loss of amplitude of the seismic signals. In fact, Eqs (2.64) and (2.65) in Section 2.7.2 show that, for particle displacement as well as stresses:

$$t_i t_i' = (1 + k_i)(1 - k_i) = 1 - k_i^2 < 1 \tag{3.29}$$

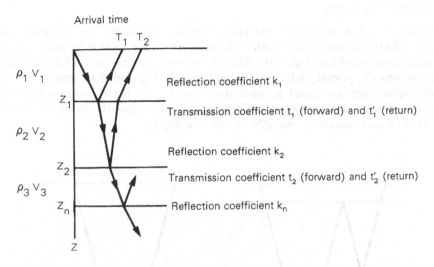

Fig. 3.6   Primary reflections.

where, for interface i:

k_i = reflection coefficient,

t_i = forward transmission coefficient (downward),

t'_i = return transmission coefficient (upward).

The seismic recording consists of a succession of arrivals of energy reflected from all the interfaces. The amplitudes are proportional to the reflection coefficients, attenuated by transmission losses. For example, the pth reflection reaches the surface at the two-way propagation time between the surface and the pth interface, and its amplitude is proportional to:

$$k_p \sum_{i=1}^{p-1} t_i t'_i = k_p \sum_{i=1}^{p-1} (1 - k_i^2) \qquad (3.30)$$

where k_p is the reflection coefficient on the pth interface, and t_i and t'_i are the downward and upward transmission coefficients across the ith interface.

b. Multiple reflections

In fact, this argument concerns only primary reflections, namely those that are reflected only once from the interfaces. In practice, in addition to primary reflections, the seismic recording contains multiple reflections that are reflected several times from the interfaces.

- "Long-path" multiples:

Multiples can be created by reflections from several interfaces at some distance from each other, with the raypaths shown in Fig. 3.7a. These are "long-path" multiples which yield spurious energy arrivals at times when no reflection should occur, or which disturb the simple reflections and alter their amplitudes. In the discussion of data processing, we shall show that processes are available for inhibiting or attenuating "long-path" multiples.

- "Short-path" multiples:

Interfaces are often numerous and close together in nature, and another type of secondary reflection exists, "short-path" multiples, which are created by reflections from two or more close interfaces (Fig. 3.7b). Theory and experience show that these multiples often strengthen the primary reflections with a certain lag, and that, unlike the "long-path" multiples, they are useful because they compensate for the attenuation of the primary reflections due to transmission losses. Without them, the primary reflections would often be undetectable, especially at extreme depths.

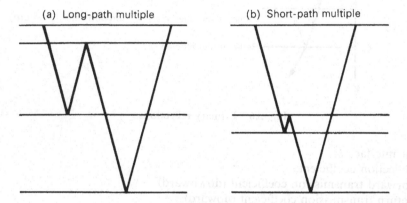

Fig. 3.7   Two types of multiple reflections.

Thus the succession of echos obtained in reflection shooting comprises a series of primary reflections and multiple reflections. The recorded seismograms are the cumulation of the primary and multiple reflections, to which noise is added.

3.6.1 Synthetic seismograms

It is theoretically possible to calculate the reflection seismogram that would be obtained on a layered half space, characterized by a superposition of layers of different densities and velocities. In the specific case in which the formations are perfectly elastic beds separated by plane and horizontal interfaces, the disturbance a plane and horizontal compressional wave, and the geophones positioned along the surface, the computation is particularly simple. Since the formation is perfectly elastic, no absorption of seismic waves occurs, and the pulse propagates in the subsurface without deformation. No conversion of P- to S-waves occurs across the interfaces, because the interfaces are parallel to the wavefronts, and only the P-wave propagation velocities and the P-wave reflection and transmission coefficients in normal incidence are significant. The problem is limited to one dimension Oz.

Note that, in the specific case of plane waves in normal incidence, the reflection and transmission coefficients of the particle displacement, on an interface between two media with densities ρ_i and ρ_{i+1} and velocities V_i and V_{i+1}, are written:

$$\text{Reflection coefficient:}\quad k_i = \frac{\rho_i V_i - \rho_{i+1} V_{i+1}}{\rho_i V_i + \rho_{i+1} V_{i+1}}$$

$$\text{Transmission coefficient:}\quad t_i = \frac{2\rho_i V_i}{\rho_i V_i + \rho_{i+1} V_{i+1}}$$

These coefficients depend exclusively on the acoustic impedances $Z = \rho V$ of the two media in contact.

For plane waves in normal incidence, two types of synthetic seismogram are often computed, the seismogram without multiples or transmission losses, and the seismogram with multiples and transmission losses.

3.6.1.1 Synthetic seismogram without multiples or transmission losses

Let us assume a velocity distribution $V(z)$ (compressional velocity, for example) and a density distribution $\rho(z)$, obtained from well logs as a function of depth. We begin by converting $V(z)$ and $\rho(z)$ into $V(T)$ and $\rho(T)$, which are functions of the vertical two-way propagation time across the formations. This operation raises no particular difficulty in so far as the velocity $V(z)$ is "calibrated" by well-velocity surveys which directly measure the propagation time between the surface and well geophones placed at successive depths.

The calibrated function $V(z)$ must be such that the integral: $\int_0^p \frac{dz}{V(z)}$ is identical to the propagation time t_p between the surface and depth p. The functions $V(T)$ and $\rho(T)$ are sampled with a sufficiently narrow interval ΔT, 1 ms for example, to represent correctly the finest significant lithological variations. This operation is performed with digitizing machines, which generate a data file containing the values of V_i and ρ_i for layers i of time-thickness ΔT.

The time series of reflection coefficients is given by the formula:

$$k_i = \frac{\rho_i V - \rho_{i+1} V_{i+1}}{\rho_i V_i + \rho_{i+1} V_{i+1}} \tag{3.31}$$

where i ranges between 1 and n, the interface i separates the layers i and $i + 1$, and n is the deepest interface. Figure 3.8 shows the time series of reflection coefficients without transmission losses and with transmission losses, sampled with an interval $\Delta T = 1$ ms. The series of coefficients with transmission losses is called the impulse synthetic seismogram h_i.

The actual synthetic seismogram without multiples or transmission losses is obtained by convolving the time series of reflection coefficients by the seismic pulse $s(t)$. Like the reflection coefficients, the pulse $s(t)$ is sampled with an interval ΔT and the synthetic seismogram without multiples y_j is given by the convolution algorithm:

$$y_j = \Delta T \sum_{i=1}^{n} k_i s_{j-i} \tag{3.32}$$

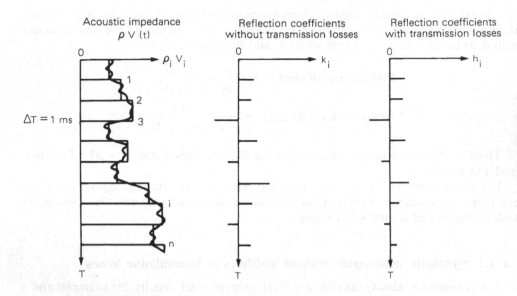

Fig. 3.8 Principle for computing synthetic seismograms: sampled acoustic impedance log, reflection coefficients k_i, and impulse seismogram h_i.

As a rule, the ΔT selected is 1, 2 or 4 ms, depending on the detail of the results desired. Synthetic seismograms without multiples or transmission losses allow prediction of the reflection seismograms that would be obtained in practice in normal incidence, in areas where the stratification is substantially tabular, after the recordings are corrected to a zero source geophone distance, and after the elimination of multiple reflections.

3.6.1.2 Synthetic seismogram with multiples and transmission losses

A more complete computation helps to obtain the synthetic seismogram with multiples while accounting for transmission losses (Baranov and Kunetz, 1960, Wuenschel, 1960, Bois *et al.*, 1960). This seismogram is the convolution of the pulse $s(t)$ by the impulse synthetic seismogram h_i:

$$Y_j = \Delta T \sum_{i=1}^{n} h_i s_{j-i} \qquad (3.33)$$

The impulse seismogram h_i is the response to a Dirac pulse of the series of reflection coefficients k_i. The entire problem consists in computing h_i from k_i.

Computation of the impulse seismogram h_i

Let w and σ be the vertical displacement and the vertical stress in each layer. It was shown in Section 2.4 that the stress σ is obtained from the displacement w by the equation:

$$\sigma = \rho V^2 \frac{\partial w}{\partial z} \qquad (3.34)$$

The displacement w is the sum of the displacement of particles corresponding to a descending wave $a(t - z/V)$ and to an ascending wave $b(t + z/V)$.

In the harmonic domain, the particle displacement w_{m_1} and the stress σ_{m_1} at the base of layer m can be expressed as a function of the amplitudes a_m and b_m of the descending and ascending waves, at the top of layer m by the equations:

$$
\begin{cases}
w_{m_1} = a_m \exp\left(j\omega \dfrac{z_{m-1} - z_m}{V_m}\right) + b_m \exp\left(-j\omega \dfrac{z_{m-1} - z_m}{V_m}\right) & (3.35) \\[2mm]
\sigma_{m_1} = \rho_m V_m \left[- a_m \exp\left(j\omega \dfrac{z_{m-1} - z_m}{V_m}\right) + b_m \exp\left(-j\omega \dfrac{z_{m-1} - z_m}{V_m}\right)\right] j\omega & (3.36)
\end{cases}
$$

The displacement w_{m_2} and stress σ_{m_2} at the top of layer $m + 1$ can be expressed as a function of the amplitude a_{m+1} and b_{m+1} at the top of layer $m + 1$ by the equations:

$$
\begin{cases}
w_{m_2} = a_{m+1} + b_{m+1} & (3.37) \\[2mm]
\sigma_{m_2} = \rho_{m+1} V_{m+1} [- a_{m+1} + b_{m+1}] j\omega & (3.38)
\end{cases}
$$

The boundary conditions on interface m express the continuity of the displacements and vertical stresses on either side of interface m:

$$
\begin{cases}
w_{m_1} = w_{m_2} & (3.39) \\[2mm]
\sigma_{m_1} = \sigma_{m_2} & (3.40)
\end{cases}
$$

in other words:

$$
\begin{cases}
a_m \exp(-j\omega\tau) + b_m \exp(j\omega\tau) = a_{m+1} + b_{m+1} & (3.41) \\[2mm]
\rho_m V_m [a_m \exp(-j\omega\tau) - b_m \exp(j\omega\tau)] = \rho_{m+1} V_{m+1}[a_{m+1} - b_{m+1}] & (3.42)
\end{cases}
$$

where τ denotes the simple propagation time across layer m, a_m and b_m, and a_{m+1} and b_{m+1} the particle motion of the descending and ascending waves at the top of layers m and $m + 1$.

The system of linear equations is conveniently written in matrix form:

$$
\begin{pmatrix} a_m \\ b_m \end{pmatrix}
\begin{pmatrix} \exp(-j\omega\tau) & \exp(j\omega\tau) \\ \exp(-j\omega\tau) & -\exp(j\omega\tau) \end{pmatrix}
=
\begin{pmatrix} 1 & 1 \\ \dfrac{Z_{m+1}}{Z_m} & -\dfrac{Z_{m+1}}{Z_m} \end{pmatrix}
\begin{pmatrix} a_{m+1} \\ b_{m+1} \end{pmatrix}
\tag{3.43}
$$

where Z_m and Z denote the acoustic impedances $\rho_m V_m$ and $\rho_{m+1} V_{m+1}$ of layers m and $m + 1$. The first square matrix depends only on the interval transit-time τ and on the angular frequency ω, and the second on the ratio of acoustic impedances in layers m and $m + 1$.

By inversing the first square matrix, we obtain:

$$
\begin{pmatrix} a_m \\ b_m \end{pmatrix}
=
\begin{pmatrix} \dfrac{\exp(j\omega\tau)}{2} & \dfrac{\exp(j\omega\tau)}{2} \\[2mm] \dfrac{\exp(-j\omega\tau)}{2} & -\dfrac{\exp(-j\omega\tau)}{2} \end{pmatrix}
\begin{pmatrix} 1 & 1 \\ \dfrac{Z_{m+1}}{Z_m} & -\dfrac{Z_{m+1}}{Z_m} \end{pmatrix}
\begin{pmatrix} a_{m+1} \\ b_{m+1} \end{pmatrix}
$$

$$
= \frac{x^{-\frac{1}{2}}}{1 + k_m}
\begin{pmatrix} 1 & k_m \\ x k_m & x \end{pmatrix}
\begin{pmatrix} a_{m+1} \\ b_{m+1} \end{pmatrix}
\tag{3.44}
$$

and by setting:

$$x = \exp(-2j\omega\tau)$$

$$k_m = \frac{Z_m - Z_{m+1}}{Z_m + Z_{m+1}}$$

the matrix equation which relates the amplitudes a_1 and b_1 of the displacements in the first layer and the amplitude a_n of the descending wave in the last layer can be written:

$$\begin{pmatrix} a_1 \\ b_1 \end{pmatrix} = x^{\frac{-(n-1)}{2}} \prod_{m=1}^{n-1} \left[\frac{1}{1+k_m} \begin{pmatrix} 1 & k_m \\ xk_m & x \end{pmatrix} \right] \begin{pmatrix} a_n \\ 0 \end{pmatrix} \tag{3.45}$$

where the amplitude b_n of the ascending wave in the last layer is assumed to be zero. (3.45) can be written:

$$\begin{pmatrix} a_1 \\ b_1 \end{pmatrix} = \begin{pmatrix} P \\ Q \end{pmatrix} a_n \tag{3.46}$$

The source can be introduced by assuming that it is located on an interface.

In the specific case in which the source and receiver are located on the surface, it can be shown (Bois *et al.*, 1962) that the spectrum of the impulse seismogram is expressed exclusively as a function of P and Q as follows:

$$R = \frac{P + Q}{P + k_0 Q} \tag{3.47}$$

where k_0 denotes the air/ground reflection coefficient (generally taken as -1 considering the particle displacement reflection coefficient). R is the response of the stack of layers to a Dirac pulse in the harmonic domain.

To obtain the impulse synthetic seismogram, the response must be computed in the time domain. Thus the inverse Fourier transform of R is obtained. Note that R can be expressed in the form of a polynomial of degree p:

$$R = \sum_p h_p x^p \quad \text{with } x = \exp(-2j\omega\tau)$$

Since the inverse Fourier transform of x is $\delta(t - 2\tau)$, the inverse transform of R has the form:

$$\sum_p h_p \delta(t - 2p\tau)$$

Hence the polynomial R immediately delivers the samples h_p of the impulse seismogram as a function of time.

If the source and receiver are not on the surface, the computation is slightly more complicated, but the principle remains unchanged (Bois *et al.*, 1962). It is mentioned in Section 4.2.13 how advances in data processing now make it possible to compute synthetic seismograms beyond the simple cases of normal incidence and horizontal curves.

3.6.2 Reflection seismograms obtained in seismic prospecting

The seismogram $Y(t)$ recorded in reflection prospecting is the result of the convolution of the seismic pulse $s(t)$ by the impulse response $h(t)$ of the subsurface, to which noise $b(t)$ is added:

$$Y(t) = h(t) * s(t) + b(t) \tag{3.48}$$

The recorded seismogram is close to the impulse response $h(t)$ of the subsurface if $s(t)$ is close to a Dirac pulse and if the noise $b(t)$ is low.

3.6.2.1 Short signals

In seismic prospecting with short signals (explosive and impact sources onshore, air-, water-, steam-guns, imploders offshore), the transmitted pulse $s(t)$ has a duration of a few tens of milliseconds, and its frequency spectrum can spread from 10 to 100 Hz in conventional seismic prospecting, or from 20 to about 200 Hz in high-resolution seismic prospecting. Figure 3.9 shows a number of pulses $s(t)$ and their frequency spectra.

3.6.2.2 Long sweep signals. Vibroseismic prospecting

In seismic prospecting with vibrators, the pulse $s(t)$ is far from short, and long signals are used, lasting several seconds covering a wide frequency band: for example, signals with a uniformly variable frequency between 10 and 70 Hz (Fig. 3.10). Seismograms recorded with long signals are evidently far from the impulse response of the subsurface. In order to separate the seismic echos, it is necessary to perform a signal compression to "gather together" the long signal in a short pulse.

a. Signal compression

This can be carried out by cross-correlation operations. Let $s(t)$ be the long sweep signal transmitted, and $Y(t)$ the signal received. The cross-correlation of the received signal and the transmitted signal is written:

$$r(t) = \int_{-\infty}^{+\infty} Y(\tau)s(t + \tau)\, d\tau \tag{3.49}$$

$$= Y(t) * s(-t) \text{ in symbolic writing}$$

By replacing $Y(t)$ by its expression (3.48) as a function of the sweep signal $s(t)$, the subsurface impulse reponse $h(t)$, and the noise, we obtain:

$$r(t) = [h(t) * s(t) + b(t)] * s(-t)$$
$$= h(t) * s(t) * s(-t) + b(t) * s(-t) \tag{3.50}$$

or

$$r(t) = h(t) * a(t) + b(t) * s(-t) \tag{3.51}$$

where $a(t) = s(t) * s(-t)$ is the autocorrelation of the sweep signal.

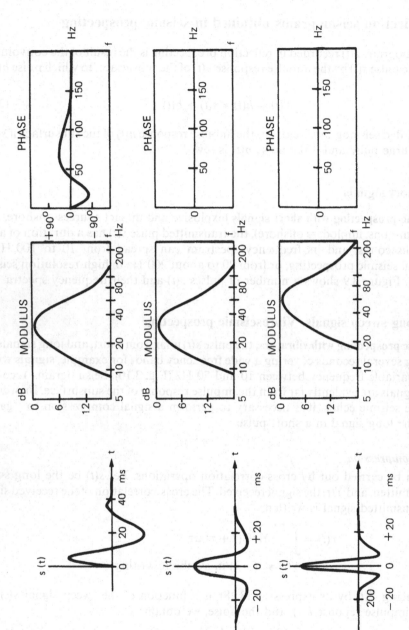

Fig. 3.9 Three pulses $s(t)$ and their frequency spectra (moduli and phases).

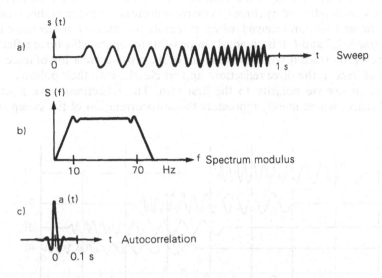

Fig. 3.10 (a) Vibroseismic sweep signal $s(t)$; (b) Spectrum modulus $|S(f)|$, and (c) Autocorrelation $a(t)$ of $s(t)$.

The cross-correlation of the seismogram received with the sweep signal thus amounts to convolving the impulse response $h(t)$ with the autocorrelation $a(t)$, and cross-correlating the noise $b(t)$ with the sweep signal $s(t)$. The final seismogram, which results from the cross-correlation, is hence an identical seismogram to the one that would have been obtained with a source delivering a signal $a(t)$, with the noise filtered by sweep signal $s(t)$.

b. Choice of the sweep signal

The choice of the sweep signal in vibroseismic prospecting is fairly flexible and allows an optimization that is not feasible with pulse sources. A sweep signal $s(t)$ is generally selected whose autocorrelation $a(t)$ is as favorable as possible with respect to the detail and the signal-to-noise ratio. The autocorrelation $a(t)$ must be short and have a high-amplitude central arch. It can be shown that this condition is satisfied if the signal $s(t)$ covers a sufficiently wide frequency band (2 to 3 octaves) and if its duration is sufficiently long (more than 1 s for example).

Figure 3.10 shows a sweep signal $s(t)$ used in vibroseismic prospecting, its spectrum $S(f)$ and its autocorrelation $a(t)$. With such a signal, the autocorrelation has a high amplitude and short duration (30 to 50 ms).

Note that the autocorrelation $a(t)$ is theoretically symmetrical, since its phase is zero. In fact, the received pulse is not identical to the transmitted pulse because of the signal deformation in the subsurface, and, in practice, $a(t)$ is not an autocorrelation, but the cross-correlation of the transmitted pulse by the received pulse. Hence $a(t)$ is dissymmetric, and it can be shown that its duration stretches for increasingly deep reflectors. There is a loss of "resolution" with depth, with both the long and short signals.

Figure 3.11 shows the principle of a Vibroseis recording. The sweep signal is the upper trace, and the signals reflected by three successive reflectors are the next three traces. The fifth trace is the seismogram received, which is merely the addition of the three reflected signals from traces 2, 3 and 4. It is obviously impossible to separate the three reflections on trace 5. However, on trace 6, which is the result of the cross-correlation of trace 5 by the sweep signal of trace 1, the three reflections appear clearly, with their polarity: the third reflection has an inverse polarity to the first two. The reflections have a short and symmetrical shape, which merely represents the autocorrelation of the sweep signal.

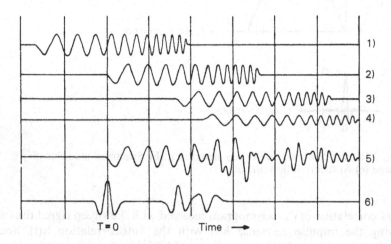

Fig. 3.11 Principle of Vibroseis record. (From Sheriff, 1973, *Encyclopedic Dictionary of Exploration Geophysics*, Society of Exploration Geophysicists, Tulsa, Oklahoma, USA).

(1) Sweep signal. (2) Return from reflector *A*; (3) Return from reflector *B*; (4) Return from reflector *C*; (5) Seismogram received before correlation, and (6) Output trace resulting from correlating (5) with (1).

3.7 RESOLUTION AND DETECTION

3.7.1 Vertical resolution

Vertical resolution is the possibility of separating two horizons at depth by reflection surveying. It depends on the dominant frequency (the reciprocal of the peak-to-peak time interval), and on the pass-band of the pulse spectrum. In fact, the dominant frequency must be sufficiently high to give a short wavelength, and the spectrum pass-band must be sufficiently broad to avoid an excessive number of oscillations. For example, let us consider a plane, horizontal wave reflecting from a thin layer. If the main wavelength Λ in the layer is much less than the thickness, and if the seismic pulse has only a small number of oscillations, the seismic recording allows an easy distinction between the top and the bottom of the layer. By contrast, if the wavelength is much greater than the thickness, or if the seismic pulse has an excessive number of oscillations, it is impossible to separate the top from the bottom. The minimum thickness of a bed in which the top and bottom can be distinguished separately is the resolution limit. Experience shows that it is about $\Lambda/4$ with the usual seismic pulses and in the presence of low noise, but that it can reach $\Lambda/2$ in the presence of higher noise.

Synthetic examples. The synthetic example in Fig. 3.12 shows the reflection on a pinchout consisting of a layer with velocity V_2 buried in a medium with velocity V_1. At places where the layer thickness h is equal to or greater than $\Lambda/4$, the top can be distinguished easily from the sole. If, however, h is less than $\Lambda/4$, only one reflection can be

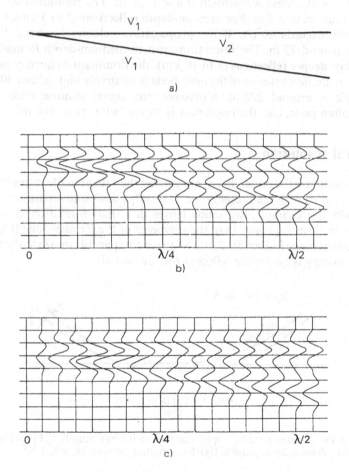

Fig. 3.12 Reflection from a pinchout immersed in a constant-velocity medium. The pinchout thickness is shown by wavelength $\left(\lambda = \dfrac{V_2}{f} \right)$. (From Sheriff 1980, *Seismic Stratigraphy*, IHRDC, Boston, Massachusetts, USA).

(a) Model; **(b)** Response to a minimum-phase pulse, and **(c)** Response to a zero-phase pulse.

distinguished, the result of the interference of the reflections from the top and sole of the pinchout; it can be seen that the interference is constructive (strengthening the reflected signal) for h between $\Lambda/10$ and $\Lambda/6$, and destructive for h less than $\Lambda/10$. The reflection becomes evanescent for h tending toward zero. The resolution is clearly around $\Lambda/4$ in the absence of noise.

Actual cases. Owing to the inelastic properties of the subsurface (Section 2.5), the seismic pulse tends to be deformed during propagation. It loses its high frequency content and stretches. This leads to a decrease in resolution with depth. For **shallow reflections** (less than 500 m), the pulse contains relatively high frequencies, and the dominant frequency may be up to 100 Hz. If the propagation velocities in the surface formations are in the range of 2000 m/s, the main wavelength is about 20 m. The resolution limit in shallow formations is thus around 5 m. For **medium-depth reflections** (2 to 3 km), the dominant frequency rarely exceeds 40 Hz. If the propagation velocity is about 3000 m/s, the wavelength is around 75 m. The resolution limit in medium-depth formations is often about 20 m. For **deeper reflections** (8 to 10 km), the dominant frequency rarely exceeds 20 Hz. The propagation velocity of the deep beds is relatively high (about 4000 m/s), and the wavelength is around 200 m. Moreover, the signal-to-noise ratio of the deep reflections is often poor, and the resolution is rarely better than 100 m.

3.7.2 Lateral resolution

Lateral resolution is the possibility of separating the subsurface features laterally. As in optics, it is associated with the Fresnel zone. It is only possible to separate two subsurface features laterally (for example, two sandy lenses in a shaly formation) if the horizontal distance between them is greater than the diameter of the Fresnel zone (Fig. 3.13). This recalls the remarks made in Section 2.9 on features comparable in size to the wavelengths, for which the energy is no longer reflected but diffracted.

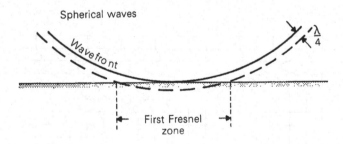

Fig. 3.13   Fresnel zone for a spherical wave with wavelength λ. (From Sheriff, 1980, *Seismic Stratigraphy*, IHRDC, Boston, Massachusetts, USA).

The lateral resolution is also limited by the space sampling, namely the distance between the geophones or between the geophone spreads summed on the same recording channel. If the space sampling is too wide (greater than 20 m in fine seismic prospecting, or 50 m in conventional seismic prospecting), ambiguities are likely to arise in the determination of high-dip beds.

3.7.3 Detection

This is the possibility of obtaining reflections from thin beds, and the detection limit is the minimum thickness of a layer visible in reflection prospecting. It depends on the wavelength, the signal-to-noise ratio, and the velocity contrast. If a bed with velocity V_2 is buried in a homogeneous medium with velocity V_1, and if the noise is low and the contrast V_2/V_1 relatively high, the detection limit may drop to values in the range of $\Lambda/30$, where Λ is the wavelength in the layer (Sheriff, 1980). **In practice,** detection is far from being this good. In stratified series and in the presence of noise, experience shows that the detection limit is around $\Lambda/4$.

3.8 THE NECESSITY OF CORRECT IMPLEMENTATION OF FIELD SURVEYS AND THE NEED FOR COMPUTER PROCESSING

The purpose of seismic prospecting is to obtain data of a structural nature (position of interfaces) and a geological nature (types of formation, facies identification). To achieve this, field surveys must be conducted with appropriate procedures, designed to strengthen reflections, attenuate noise, and enhance the accuracy of the results. These techniques, multiple coverage, geophone and source patterns, three-dimensional implementation, have become feasible with multichannel digital recording (12, 24, 48 and 96, and now several hundred channels) thanks to the increasing storage and processing capabilities of computers.

CHAPTER

4

reflection surveys

Chapters 4 to 6 deal chiefly with data acquisition and processing in reflection and refraction surveys. A few pages are devoted to transmission surveys and well surveys, with some emphasis on vertical seismic profiles.

Reflection seismic surveying is an exploration method in which disturbances are generated in the ground, and the waves reflected from the geological interfaces are observed on the surface. This helps to map subsurface structures, by measuring the arrival times of the reflected events, and to determine stratigraphic features by analyzing the characteristics of the reflected signals.

4.1 DATA ACQUISITION

4.1.1 Acquisition systems

Acquisition systems essentially comprise a source pattern, a detection spread, and digital recording instruments. They are different in land and marine surveys.

a. Land surveys

The acquisition principle initially used was that of **single-fold implementation.** In the example shown in Fig. 4.1, a detection spread with 24 seismic channels is connected to 24 geophones at 50 m intervals. Emission is provided by a source S_1 located at the center of the spread, and the portion of reflector "covered" by the seismic raypaths is equal to one half-spread. For a 1100 m detection spread, for example, the portion of reflector M_1 covered by the seismic beam emitted by S_1 is 550 m. Geophones 1 to 12 are then moved to 1' to 12', and an emission is generated at S_2, at the center of the new spread. The portion of reflector M_2 covered by the beam emitted by S_2 extends over 550 m from M_1. Geophones

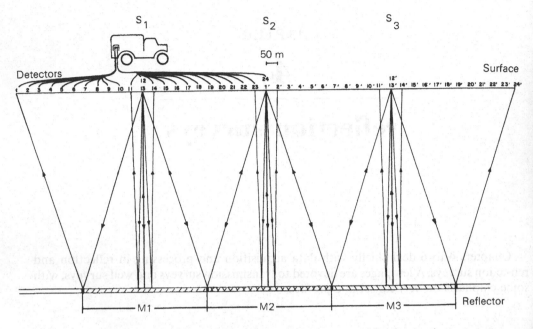

Fig. 4.1 Acquisition spread in single-fold coverage (land survey).

13 to 24 are then moved to 13′ to 24′, and an emission is generated at S_3, to cover M_3, and, so on. By successively moving the source/detector system by a half-spread length, the entire reflector is covered once. Hence the name of "single-fold implementation" given to this method.

The acquisition principle generally employed today is not that of single-fold, but that of **multifold implementation** (Waters, 1978). The source is not moved a half-spread, but a shorter distance. By shifting the source 1/12 or 1/24 of a spread, for example, the reflectors can be "covered" 6 or 12 times by the same "common midpoints" (CMP). If the source shift is equal to the trace interval, the result is called a "roll-along" spread, and the CMP coverage is equal to half the number of traces. For a 24-trace-spread, for example, the roll-along method gives a 12-fold CMP coverage (Fig. 4.2). Land shooting often makes use of 96-trace spreads with a 25 m interval between traces and sources. These roll-along spreads give a 48-fold CMP coverage.

To attenuate surface noise, ten or more geophones are grouped on each seismic channel in a pattern (see Section 3.5). The geophones are lined up along the profile and each geophone group is spread over a pattern of several dozen meters, the position of the seismic trace being the center of the group.

b. Marine surveys

Marine surveys often make use of 96-trace spreads extending over 2400 m, with patterns of several dozen hydrophones per trace over more than 20 m (Fig. 4.3). The traces usually have a 25 m spacing, and all the hydrophones are placed in a 2400 m long flexible hose (the streamer) towed by the ship at a speed of about 4 knots (2 m/s). The source array

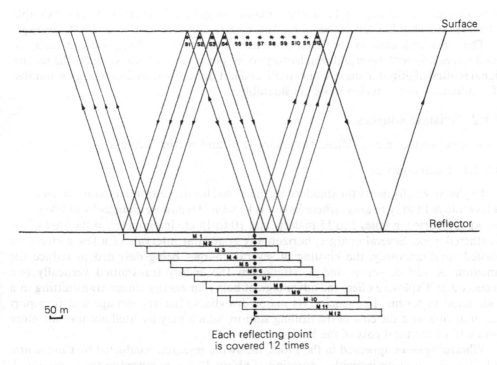

Fig. 4.2   Acquisition spread in multifold CMP coverage ("roll-along" pattern 12-fold CMP coverage, land survey).

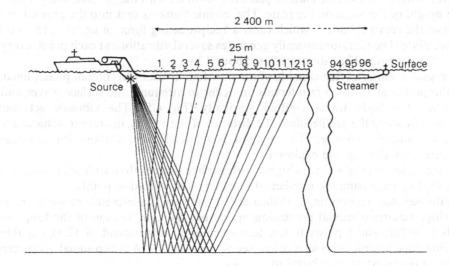

Fig. 4.3   Acquisition spread in marine survey.

is also towed by the ship, and a seismic emission is generated every 12.5 s, for example, whenever the ship has advanced 25 m. This clearly achieves a 48-fold CMP coverage.

The streamers used in the latest systems are up to 3 km long, with as many as 1000 traces and with several dozen hydrophones per trace. In these systems, the seismic signal is often digitized in the streamer itself, and not onboard, thus decreasing the number of conductors and reducing noise considerably.

4.1.2 Seismic sources

Seismic sources are of different types for land and marine surveys.

4.1.2.1 Land sources

Explosive was formerly the standard method used for the seismic emission on land. It is still employed today in areas where its use is feasible. Dynamite charges from 100 g to a few kg are buried in holes 3 to 15 m deep, and 10 to 15 cm in diameter, at the base of the weathered zone. Several charges, horizontally spaced at intervals of a few meters, are blasted simultaneously, the blasting of several charges being designed to reduce the emission of surface waves and to strengthen the energy transmitted vertically (see Section 3.5). Explosive offers the advantage of being an energy source transmitting in a wide band of seismic frequencies. Its major drawbacks are the storage and transport requirements, and the cost of the drilling station, which may by itself account for more than half of the total cost of the land survey.

Vibrator sources appeared in the 1960s, following research conducted by *Continental Oil Company*. Wavetrains with a duration of about 12 s are emitted in the ground with frequency varying progressively between 10 and 70 Hz, for example. This is the "sweep". Vibrator trucks are used mainly, equipped with a vibrator plate actuated by a hydraulic servo cylinder system (Fig. 4.4). When the truck reaches the vibration position, the operator lowers the vibrator plate to place it in contact with the ground, and presses the entire weight of the truck on the plate. The seismic signal is sent into the ground by the action of the servo cylinders, which exert a reciprocating force of up to ± 15 t on the vibrator plate. The vibrator normally generates several vibrations at each point, every 15 or 20 s for example. The vibrations last about 12 s.

The source pattern often consists of three or four trucks vibrating in phase, lined up with the profile at 10 to 15 m intervals, in order to attenuate the surface waves and to strengthen the body waves emitted vertically (Fig. 4.5). The vibrator set moves progressively along the profile, often at a distance of 10 to 15 m, in order to achieve a high degree of multiple coverage. The method is relatively fast, without the constraints associated with drilling and explosives.

Surface noise, which is always higher with surface sources than with buried sources, is attenuated by increasing the number of vibrators and vibration points.

"Vibroseismic prospecting" (Vibroseis[1]), a very widespread process in land surveying, requires a special processing operation, the compression of the long sweep signals to obtain short pulses (a few tens of milliseconds instead of 12 s), capable of achieving good seismic echo separation (see Section 3.6.2.2). A sweep signal, its spectrum and its autocorrelation are shown in Fig. 4.6.

---

(1) *Continental Oil Company* trademark.

Fig. 4.4   Mertz M13 vibrator truck. (*Mertz* Document).

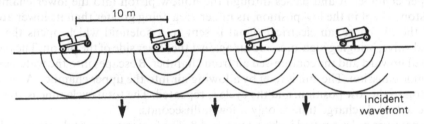

Fig. 4.5   Conventional source pattern in vibroseismic land surveying. Four vibrator trucks in phase.

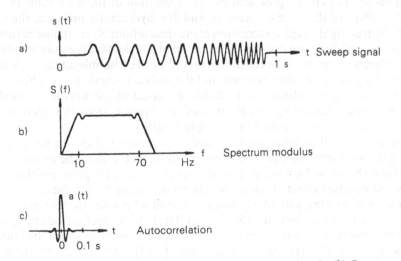

Fig. 4.6   Example of long sweep signal. **(a)** Sweep signal; **(b)** Spectrum modulus, and **(c)** Autocorrelation.

Other land sources are simple impactors (weight drop on a target like the IFP Soursile) and coded impactors (mechanical tampers striking the ground at a controlled frequency, like the SNEAP Minisosie). They are often employed for fine seismic investigations with high resolution and medium penetration. Other systems employ metal plates thrust against the ground by the explosion of a gas mixture (Atlantic Richfield Dinoseis) or a compressed air discharge into small water tanks placed on the ground surface (Bolt air guns).

4.1.2.2 Marine sources

Explosive, formerly used for offshore seismic emissions, has nearly completely disappeared today, mainly because of the damage it often inflicted on the marine fauna.

Air guns are the most widely used sources today. The principle consists of the injection of a few liters of highly-compressed air into the water, causing the emission of a high-intensity pressure wave (McQuillin *et al.*, 1979). Figure 4.7 illustrates the operating principle of air guns. The instrument consists of two high-pressure chambers A and B closed by a piston. During the loading period, the high-pressure air (140 to 200 bar) enters the upper chamber A and passes through the hollow piston into the lower chamber B. The piston is kept in the low position, its upper area being greater than its lower area. To trigger the discharge, an electrical signal is sent to a solenoid which opens the upper solenoid valve, bringing high-pressure air against the lower side of the piston. The piston is propelled upward and the compressed air from chamber B escapes into the water, causing the seismic emission. The gun is reset by allowing air into the upper chamber A, returning the piston to the low position, and the cycle is repeated. The total cycle time is 10 to 15 s, and the actual discharge time is only a few milliseconds.

The pressure pulse emitted is shown in Fig. 4.8. The first pressure peak, measured in the 0 to 250 Hz frequency band, exceeds 2 bar at 1 m, with a 1.64 liter (100 cu. in.) gun fired at a depth of 6 m. This first pressure peak is followed by several secondary emissions, at 90 and 180 ms for 100 cu. in., produced by the oscillation of the air bubble in the water. Under the effect of the internal pressure and the hydrostatic pressure, the air bubble oscillates by passing through a succession of maxima (about 50 cm diameter) and minima (a few cm), on each occasion causing the emission of a secondary pressure wave at the time of the minimum diameter. The repetition period of the "bubble effect" depends on the volume of the gun, the loading pressure and the emission depth. For a 100 cu. in. gun at 280 bar loading pressure (4000 psi) at a depth of 6 m (20 ft), the repetition period is 90 ms. All other things remaining equal, it can be shown that this period increases approximatively as the cube root of the stored energy.

A convenient method often used to inhibit secondary pulses consists in positioning several guns of different volumes a few meters from each other. The initial pulses are in phase, while the secondary pulses are not, so that the total pulse emitted downward displays a strengthened initial pulse and attenuated secondary pulses.

Figure 4.8 shows the individual pulses obtained with four different guns, ranging in volume from 0.33 to 1.64 liter (20 to 80 cu. in.). The bubble effect periods range from 60 to 90 ms. The combination of different guns gives a single signal, in which the initial pulse is strengthened and the secondary pulses are highly attenuated. Modern emission techniques use several gun arrays, spread over several dozen meters and shooting

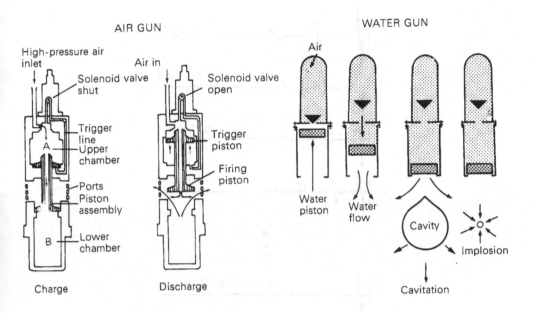

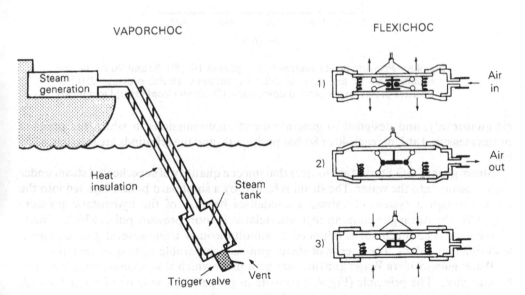

Fig. 4.7 Marine seismic sources. Air gun (Bolt PAR), Vaporchoc (CGG), Water gun (Sodera), Flexichoc (IFP). (From R. McQuillin, M. Bacon and W. Barclay, 1979. *An Introduction to Seismic Interpretation*, published by Graham and Trotman, London).

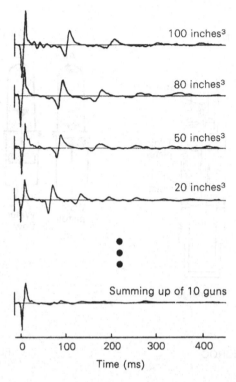

Fig. 4.8 Pressure pulses emitted by air guns of 100, 80, 50 and 20 cu. in. and summing up of ten guns of different volumes. Initial pressure 280 bar. Positive pressures are plotted downward. (*Western Geophysical* Document).

simultaneously, and designed to generate directional emissions in which the pressure pulses measured at 1 m reach 40 or 80 bar peak/peak, and the rebounds are less than a few bars.

Steam guns (CGG Starjet) are sources that inject a quantity of superheated steam under high pressure into the water. The steam is heated by a shipboard boiler. Injected into the water through a system of valves, it condenses because of the hydrostatic pressure (Fig. 4.7). The steam gun emits an energetic, relatively short pressure pulse, which exhibits a precursor. The precursor is inhibited by simultaneously using several guns equipped with different valves. The energy of steam guns is comparable to that of air guns.

Water guns (Sodera Water gun) are marine sources which also generate pulses without bubble effect. The principle (Fig. 4.7) consists in propelling a volume of water through vents into the water in a very short time. At the time of shooting, a "slug" of a few liters of water is propelled into the water mass by a piston actuated by a system of valves similar to that of air guns. After the piston reaches the end of a stroke, a negative-pressure zone is created behind the slug and a pressure pulse is emitted by implosion. The pulse emitted is highly reproducible, without bubble effect. Several water guns can be grouped, a dozen 6.5 liter (400 cu. in.) guns for example, to obtain directional effects, with energies comparable to those of air gun arrays. A large number of measurements taken by *Institut*

Français du Pétrole (IFP) have also shown that the emission spectrum is rich in high frequencies, giving water guns higher resolution without decreasing the penetration depth.

Implosive energy sources generate pulses without bubble effect. The **Flexichoc** (Fig. 4.7) developed by *IFP*, consists of two plates separated by a chamber, from which the air is pumped out. At the time of the shot, the plates are unlocked and propelled against each other by the hydrostatic pressure, creating a shockwave in the water. The emitted pulse is very short (a few milliseconds), rich in high frequencies, and allows high resolution while preserving satisfactory penetration.

Sparkers emit acoustic energy by an electrical discharge in the sea-water. Capacitor batteries, charged by a generator, are discharged on command in electrode arrays towed in the water. The energy released at each discharge is generally a few kilojoules or a few tens of kilojoules, a few per cent of the energy of air guns. It nevertheless allows the penetration of several hundred meters into the seafloor. Pulses without bubble effect can be obtained by adjusting the number of electrodes and their spacing to the sparker energy. Sparkers are ideal sources for the fine analysis of shallow horizons.

4.1.2.3 Source directivity

a. *Single source*

Directivity is the preferential direction of propagation of the seismic energy emitted by the source. It is often represented by a diagram, in polar coordinates, of the relative intensity of the seismic wave versus the propagation direction. For a **point source** exerting a **vertical force** at the ground surface, the directivity diagram is shown in Fig. 4.9, for *P*-waves and for *S*-waves. For *P*-waves, the emitted intensity is a maximum in the vertical direction and zero in the horizontal direction. For *S*-waves, it is a maximum around a direction close to 45°, and zero vertically and horizontally. This type of emission corresponds, for example, to the vertical vibrator, which is a good source for *P*-wave reflection surveying. The directivity diagram displays a symmetry of revolution about the vertical axis of the source. The minimum intensity for u_θ close to 35° is difficult to observe in the field. For a **horizontal *SH* point source,** exerting a horizontal force perpendicular to the profile at the ground surface, it can be shown (White, 1965) that the emitted *P*-wave intensity is theoretically zero in the vertical plane of the profile. The emitted *SH*-wave intensity perpendicular to the profile is high, in the vertical and horizontal directions. For an *SV* source, the emitted *SV*-wave intensity in the plane of the profile is high in the vertical direction, to about 35°, and practically zero in the horizontal direction. This type of *SV*- and *SH*-wave emission corresponds, for example, to a vibrator whose vibrator plate exerts a horizontal tangential force (shear wave vibrator), or to the "Marthor", an impact source developed by *IFP* and exerting high energy shear pulses at the ground surface.

b. *Multiple sources*

Independent of the directivity of individual sources, aligned or laterally spread source arrays have a natural directivity. The arrays examined in Section 3.5 help to attenuate surface noise and to strengthen the waves transmitted downward.

4.1.3 Seismic detectors

Seismic detectors designed to transform seismic energy into electrical voltage are generally geophones on land and hydrophones at sea.

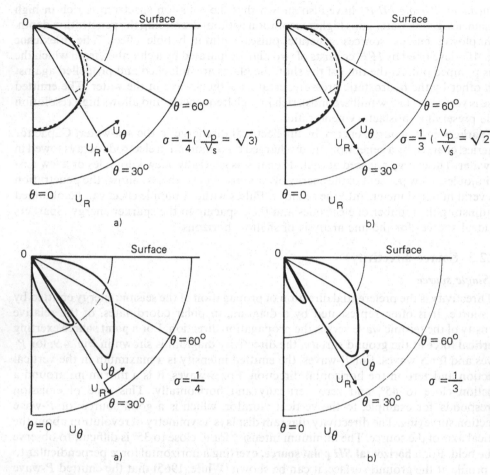

Fig. 4.9 Directivity diagram of a vertical point source on land for two different values of Poisson's ratio. (From Waters 1978, *Reflection Seismology. A Tool for Energy Resource Exploration*, John Wiley and Sons Inc., New York).

(a) $\sigma = \frac{1}{4}$, and (b) $\sigma = \frac{1}{3}$,

U_R = radial displacement of particles (*P*-waves),
U_θ = tangential displacement of particles (*S*-waves).

4.1.3.1 Geophones

Geophones transform the movement of the ground into an electrical voltage. They are usually moving coil electromagnetic sensors (Fig. 4.10). A permanent magnet in the form of a cylinder, with a radial magnetic field, contains a cylindrical slit separating the south pole (central part) from the north pole (annular part). A moving coil with a large number of very fine wire turns is suspended in the slit by light, flat springs. With the geophone placed

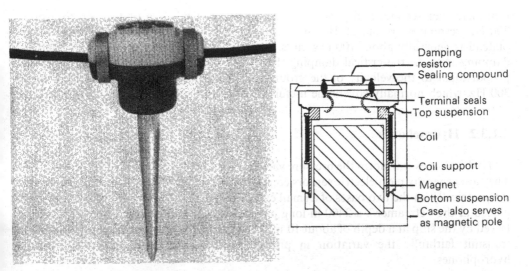

Fig. 4.10 Conventional geophone for reflection surveys (Geospace) and cross-section of a conventional moving coil, radial field geophone. (From Evenden *et al.*, 1971, *Seismic Prospecting Instruments*, Vol. 2, *Instrument Performance and Testing*, published by Gebrüder Borntraeger, Berlin).

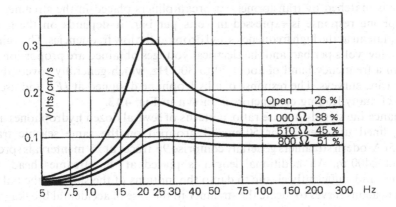

Fig. 4.11 Amplitude frequency response curves of a conventional geophone for reflection surveys. (From Evenden *et al.*, 1971, *Seismic Prospecting Instruments*, Vol. 2, *Instrument Performance and Testing*, published by Gebrüder Borntraeger, Berlin).

on the ground, the magnet follows the vertical movements of the ground, and the coil tends to remain stationary by inertia. The relative motion between the coil and the magnet generates an electrical voltage proportional to the relative velocity of the coil with respect to the magnet, to the number of turns, to the coil radius, and to the intensity of the magnetic field in the air gap. The geophone's "response" is expressed in volts per centimeter per second. It depends on the frequency, displays a plateau up to around 300 Hz, resonates around 20 Hz, and drops at the low frequencies. The value of the plateau is generally about 100 mV/cm/s, and a shunt resistance is selected to achieve damping close to the critical damping (Fig. 4.11). This produces an electrical voltage proportional to the velocity of the ground particles, in a frequency band from 20 to 300 Hz, which normally covers the useful band in land surveys.

4.1.3.2 Hydrophones

Hydrophones transform the pressure variations in the water into an electrical voltage. They are generally piezoelectric sensors consisting of two piezoelectric ceramics mounted symmetrically on the two bases of a small cylindrical drum (Fig. 4.12). Hydrophones are placed in the "streamer", a 2400 m long pipe 5 to 8 cm in diameter, filled with oil and towed by the ship at a depth of about 10 m. The streamer walls are sufficiently flexible to transmit faithfully the variation in pressure from water through the oil to the hydrophones.

The sensors are virtually unaffected by streamer acceleration effects, which cause dissymmetric deformations in the ceramics, but they are sensitive to variations in pressure which are exerted symmetrically on the two ceramics. Sensors are generally designed to compensate for the static pressure, so that their sensitivity is independent of the water depth at which they are placed. The very high impedance of the ceramic materials must be lowered to a few hundred ohms to be comparable to that of the connecting lines. The impedance is matched by transformers or preamplifiers placed in the streamer.

Hydrophone response is expressed in volts per bar. It depends on the frequency, displays a plateau at the high frequencies and drops at the low frequencies. The value of the plateau is a few volts per bar, and the electrical voltages obtained are proportional to the pressure, in a frequency band of about 20 to 300 Hz, which generally covers the useful band in marine surveys. The response of a hydrophone designed at *IFP* and used in the streamers of many leading contractors is shown in Fig. 4.13.

To enhance the signal-to-noise ratio, patterns of several dozen hydrophones are often employed, lined up over 25 or 50 m and grouped on the same seismic trace (see Section 3.5). Modern streamers generally comprise 96 traces at 25 m intervals spread over a length of 2400 m. An additional length is placed at the streamer head, without hydrophone, and sufficiently elastic to damp the motions of the ship. At the tail, a buoy fitted with beacons serves to locate the streamer in the case of accidental breakage. Depth controllers are positioned at regular intervals along the streamer, and help to keep it at constant depth (for example 10 m), and directional detection systems are provided to measure the heading and transmit it to the ship. A typical system has six depth controllers and six directional detection systems (one every 400 m).

A shipboard minicomputer computes the exact position of the streamer in the water at all times, giving the position of the subsurface reflection points investigated by the seismic raypaths for each shotpoint. The current trend is to increase the number of traces. The latest streamers have 240, 480 or even 1000 traces, with digitization in electronic modules placed in the streamer. They may be as much as 3000 m long. Only digitization in the streamer allows the recording of such a large number of traces with streamers of limited diameters. It also allows considerable enhancement of the signal-to-noise ratio.

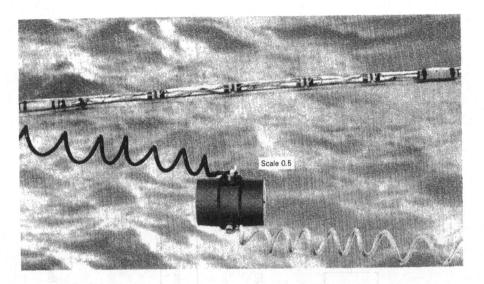

Fig. 4.12 Conventional hydrophone for reflection surveys (scale 0.5) and streamer (scale 0.05). (*Technip Géoproduction* Document).

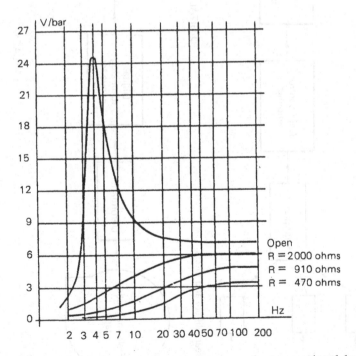

Fig. 4.13 Amplitude frequency response curves of a conventional hydrophone for reflection surveys (48 hydrophones in three parallel series). (*IFP* Document).

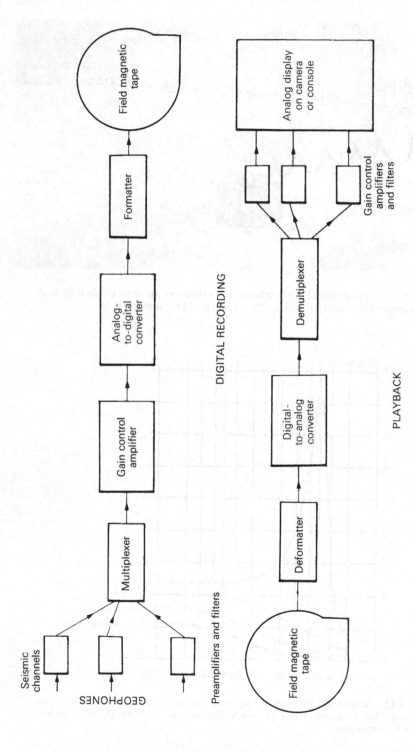

Fig. 4.14 Digital recording and field playback. Low-level multiplexing. Low-level multiplexer.

4.1.4 Digital recording

The digital recording of seismic signals appeared around 1965 and was generalized in the 1970s. As opposed to analog recording, in which the data are represented as continuous variations of a signal, digital recording represents the data as a discontinuous succession of discrete values (Pieuchot, 1978).

4.1.4.1 Digital recording system

A digital recording system (Figs 4.14 and 4.15) consists essentially of preamplifiers and analog filters, a multiplexer, a gain control amplifier, an analog-to-digital converter, a formatter, a tape unit, and a playback system.

The **preamplifiers and analog filters** help to adjust the recorded passband to the frequencies of the seismic signal, and to eliminate noise as much as possible. A low-cut filter is generally applied at about 8 Hz or so to eliminate low-frequency noise (surface waves in land surveys, streamer noise in marine surveys), and a high-cut filter at 62.5, 125 or 250 Hz to prevent aliasing. **Aliasing** is an ambiguity in the frequency resulting from the signal sampling causing frequency folding. If we consider an analog function with a spectrum like the one in Fig. 4.16 a, and if it is sampled in time with an interval $\tau = 1/f_e$, the spectrum of the sampled function becomes periodic with a period f_e (Fig. 4.16 b). It can be shown that no aliasing occurs in arrangements designed to avoid any signal at frequencies above the Nyquist frequency (half of the sampling frequency). The high-cut filter must therefore be selected accordingly. The cutoff frequency is often a quarter of the sampling frequency, with a relatively steep slope, thus eliminating any risk of aliasing. If, for example, the sampling interval is 4 ms (sampling frequency $f_e = 250$ Hz), the high-cut filter is generally selected with a cutoff frequency of 62.5 Hz and a slope of 72 dB per octave. This applies to data recording in conventional seismic surveys. In high-resolution seismic surveys, in order to record the highest frequencies, the sampling interval selected is 2 ms or 1 ms, with a high-cut filter of 125 or 250 Hz.

The **multiplexer** switches the different seismic channels in sequence at regular intervals, measures their amplitudes and feed them into a single output channel. It leaves sufficient time between each switching (a few microseconds) to measure the level of the seismic signal. Each channel is thus sampled at the sampling interval, for example every 4 ms. In low-level multiplexing techniques, which tend to become generalized, the multiplexer is placed upstream, before the gain control amplifier.

The **gain control amplifier** keeps the data transmitted to the analog-to-digital converter at an appropriate level, so that the recording remains sufficiently accurate. It is a sensitivity matcher, whose gain varies during the recording.

The variation in amplitude of the signals received by the geophones after the disturbance generally has the form shown in Fig. 4.17. The initial level, very low before the first arrival (a few microvolts), corresponds to ambient noise, to the agitation of the geophones in the wind, to automobile traffic, and to streamer tow noise in marine shooting. The level suddenly rises by a few hundred millivolts upon the arrival of the first seismic signal, and then decreases progressively, with temporary rises upon the arrival of each of the reflections. After 6 s of recording, when the deepest reflections arrive, the average level has generally dropped to a few microvolts.

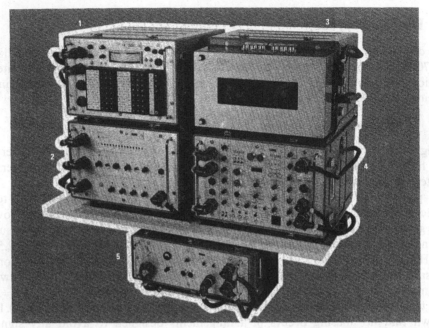

Low-level multiplexer. Instantaneous floating point. 24, 48 or 96 channels.

Fig. 4.15 SN 338 HR seismic recording instruments. (*Sercel* Document).
(1) Input and test unit; (2) Amplifier and converter; (3) Recorder; (4) Logic
unit; and (5) Power supply.

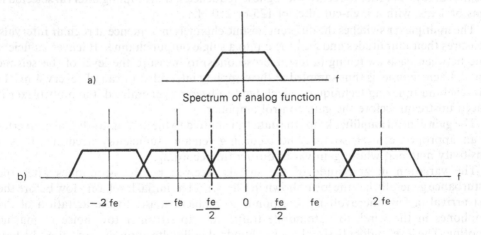

Fig. 4.16 (a) Spectrum of a continuous function, and (b) Spectrum of the
same sampled function.

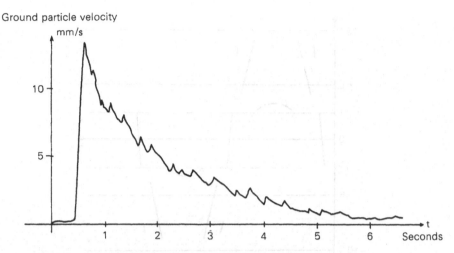

Fig. 4.17 Variation in amplitude of seismic signals in land shooting. Disturbance at time zero. Source/geophone offset 460 m.

The digital acquisition system must be able to record all these signals with a satisfactory dynamic range, in other words with maximum to minimum reading sufficient to be able to account for the strongest as well as the weakest signals. The effective dynamic range of recorders is generally inadequate to achieve this. For example, to record signals of a few microvolts simultaneously with signals of a few hundred millivolts (100 dB higher) with an accuracy of 40 dB, it requires a dynamic range of 140 dB, which is difficult to obtain with standard recorders. The recording gain is therefore adjusted differently upon the arrival of the strong signals and the arrival of the weak signals, to adjust the amplitude of the seismic signal to the dynamic range of the recorder at all times.

In binary gain control amplifiers, the reference voltages vary by a factor of 2 (6 dB). In instantaneous floating-point recording, the gain is instantaneously adjusted, so that the voltage remains as close as possible to the recorder full-scale voltage (Fig. 4.18). The output voltage V of the gain control amplifier, and the gain G with which this voltage is obtained, are encoded. Since V and G are known, the input voltage E of the seismic signal can be determined by the formula:

$$E = \frac{V}{G}$$

where the gain G is generally encoded in the form $G = 2^n$.

The **analog-to-digital converter** converts the signal into a binary code number at the output of the gain control amplifier. The binary number generally contains 16 bits, for example 12 mantissa bits and 4 gain bits. The first bit is often a sign bit. To avoid the sign bit, a constant quantity can be added to all the signals to make them all positive.

The **formatter** is a set of logic circuits designed to format the digitized information before recording on the magnetic tape. A fairly large number of possible formats exists, and it is obviously necessary to know the writing format to be able to read the magnetic tape.

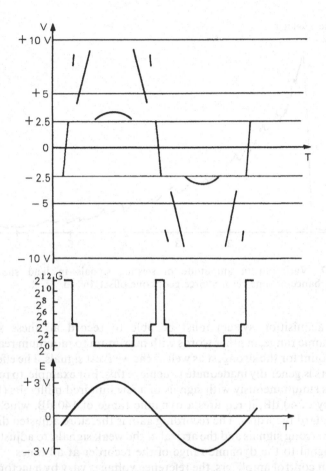

Fig. 4.18 Instantaneous floating-point representation of a sinusoid. (From Pieuchot, 1978, *Les appareillages numériques d'acquisition des données sismiques*, Éditions Technip, Paris).

The **tape unit** records the data on magnetic tape. The tapes generally measure half an inch in width, have 9 tracks, and record at 1600 or 6250 bytes per inch, at speeds ranging from 40 to 120 in/s. The first track is often used for a parity check.

Playback of magnetic tapes serves to check the quality of the recordings obtained, by giving an analog representation on camera or on console. The playback system comprises a "deformatter", which reconstructs the binary data from the code recorded on the tape, a digital-to-analog converter which generates an analog voltage from the binary data, a demultiplexer which distributes the analog voltage to the different playback channels, and a passband filter system (see Fig. 4.14).

For vibroseismic recordings, the recording truck is equipped with a correlator system, which serves to cross-correlate the received signal with the sweep signal to obtain recordings in which the reflected signals are sufficiently short, 20 to 50 ms for example, to obtain good resolution. Certain modern systems, like the BEICIP/IFP Proseis 9600, help in marine surveys to perform a preliminary processing of the recordings on line, namely at the same speed as the shooting sequence, and to obtain a shipboard seismic section processed by multiple coverage, without any prejudice to the possibilities of subsequent processing on the central computer.

4.1.4.2 Source control

Land surveys. In vibroseismic prospecting, the sweep signal is controlled by a radio signal sent from the recording truck. The operator presses the firing button to trigger the vibrations of the vibration trucks. In explosive shooting, for safety reasons firing is controlled by the shooter, who remains in permanent telephone communication with the recording truck.

Marine surveys. Acquisition is continuous and automated. A shipboard computer, which also controls the navigation, triggers the seismic disturbance at predetermined positions.

Onshore and offshore, the shot command starts up the recording lab. The recording time after each shot is about 6 s with impulse sources, and 20 s with vibration sources.

4.1.4.3 Telemetry recording systems

Recent years have witnessed the development of recording systems with a high number of channels (several hundred to one thousand channels), for the reasons listed in Section 4.1.1: enhanced resolution, noise inhibition, three-dimensional acquisition. However, it is difficult to use cable links for several hundred channels with standard systems. Telemetry systems are therefore employed, consisting of a number of digitizing units distributed along the spread, which carry out multiplexing and analog-to-digital conversion, near the geophones. In land surveys, the digitizing units are sometimes connected by radio to the recording truck, and all the operations are monitored by the central processor, as in the IFP/CGG Myriaseis system. These systems require no cable connection, thus avoiding the electrical disturbances such as industrial 60 Hz (Fig. 4.19).

Optical fiber cable connections appeared recently on the market, facilitating the elimination of electrical disturbances while allowing the transmission of a very large density of data.

4.1.5 Positioning at sea

The positions of the source and detection points in seismic prospecting must be known accurately to locate the subsurface reflectors. Standard topographic methods are employed in land surveys, generally yielding sufficient accuracy. The problem is much more complex in marine surveys, and to determine the position of the geophysical ship, radio positioning or satellite positioning methods are necessary.

4.1.5.1 Radio positioning methods

These methods can be used near the coast and in areas provided with offshore platforms. Their principle consists in measuring the distance between the ship and several fixed stations, or the difference in distance between the ship and pairs of fixed stations. In the former case (Fig. 4.20 a), distance measuring systems are employed, by installing

Fig. 4.19 Myriaseis telemetry recording system. (*IFP/CGG* Document).

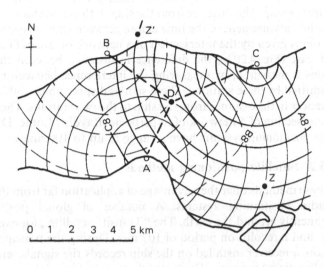

(a) Location by measurement of distances at three points. (From R. McQuillin, M. Bacon and W. Barclay, 1979, *An Introduction to Seismic Interpretation*, published by Graham and Trotman, London).

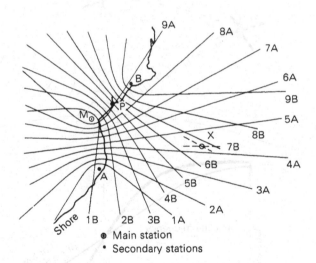

⊚ Main station
• Secondary stations

(b) Location by measurement of differences in distances from several pairs of points. (From Dobrin, 1976, *Introduction to Geophysical Prospecting*, McGraw-Hill, New York).

Fig. 4.20 Radio positioning systems.

(a) By intersections of circle networks, and **(b)** By intersection of hyperbola networks.

beacons at the fixed stations, which retransmit a radio signal upon reception of the signal sent by the ship. The distance from the ship to the fixed stations is determined on the ship from the measurement of the time elapse between transmission and reception. The ship's position is given by the intersection of a network of circles (Tellurometer system). In the latter case (Fig. 4.20 b), the difference in distance between the ship and pairs of fixed stations is measured, by measuring the interference between the fixed frequency signals transmitted by these stations. These signals display phase differences proportional to the differences in the distance from the ship to the two stations. The ship's position is given by the intersection of a network of hyperbolas (Raydist, Lorac, Decca, Toran systems, etc.). Radio positioning systems have a range of up to 100 miles.

4.1.5.2 Satellite positioning methods

These methods offer the advantage of application far from the coast, beyond the range of radio positioning systems. A number of global positioning satellites revolve permanently around the earth. The "Transit" satellites, for example, have circular polar orbits and a revolution period of 107 min. During the passage of the satellite above the horizon, a receiver installed on the ship records the signals sent by the satellite using an omnidirectional antenna. These signals contain precise data on the orbit followed by the

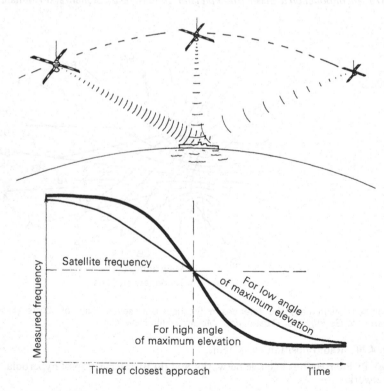

Fig. 4.21 Satellite navigation. (From Sheriff, 1973, *Encyclopedic Dictionary of Exploration Geophysics*, Society of Exploration Geophysicists, Tulsa, Oklahoma, USA).

satellite, as well as an ultrastable frequency of 150 or 400 MHz. The measurement of the Doppler effect observed on this frequency supplies data on the relative position of the satellite and the ship, the frequency received being higher as the satellite approaches, and lower as it recedes (Fig. 4.21). Moreover, the frequency variation law as a function of time serves to calculate the distance from the ship to the satellite's closest approach point. To obtain accurate results, the speed and heading of the ship must be known throughout the period of reception from the satellite (2 min). A shipboard computer integrates the satellite data, the data from a Doppler sonar (which gives the speed of the ship), and the data from a gyro-compass (which gives the heading of the ship), and possibly the data from a radio positioning system, to calculate the ship's position at any time.

The Doppler sonar (Fig. 4.22) measures the ship's velocity with respect to the seafloor, by emitting sound pulses in four narrow beams: fore, aft, port and starboard. The sound pulses are backscattered by the seafloor and the Doppler shift measurement between the fore and aft signals serves to calculate the ship's velocity in the heading direction. Comparison of the signals received at port and starboard helps to calculate the lateral drift. The Doppler sonar operates correctly up to depths of a few hundred meters. In deeper water, the speed is estimated by other systems (Doppler sonar with signal reflection from the water mass, loch systems).

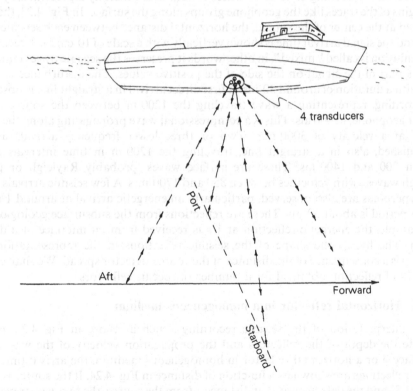

Fig. 4.22 Doppler sonar navigation principle. (From Sheriff, 1973, *Encyclopedic Dictionary of Exploration Geophysics*, Society of Exploration Geophysicists, Tulsa, Oklahoma, USA).

As a rule, the integration of all these navigation systems and the use of shipboard computers in real time help to locate the ship with absolute accuracy of about 50 m or so. This accuracy is generally sufficient to obtain satisfactory positioning of the seismic profiles. The shipboard computer controls the navigation, regularly correcting the heading to follow the profile, and checks the seismic emission cycle, triggering the shots to the precalculated shotpoints.

4.1.6 Nature of the signals received in reflection surveying

Although the subsurface structures are generally three-dimensional, two-dimensional sections are often considered satisfactory for cost reasons, with profiles in directions roughly perpendicular to the axis of the structures. Unfortunately, quite often, the subsurface geometry is such that the seismic raypaths leave the vertical plane of the profile and the seismic sections obtained are erroneous. It is then necessary to proceed with three-dimensional surveys by using the techniques described in Section 4.1.7.

In 2D as well as 3D, the signals are received by detectors generally consisting of geophone or hydrophone groups, each group connected on a single seismic channel (see Section 3.5.2). It is possible, for example, to use 48 or 96 groups (of 18 geophones), spread over 2400 m. A recording of 48 or 96 traces is obtained, generally represented by placing the origins of the traces like the geophone groups along the surface. In Fig. 4.23, the source is located at the center of the spread, the horizontal distance between each seismic trace is 50 m, and the signal arrival time is shown vertically with a scale of 10 cm for 1 second. The representation is called "mixed", in other words it preserves the seismic traces entirely but darkens part of the signal on the side of the positive values. The disturbance is a single pulse with a duration of around 30 ms. A first direct arrival in a straight line is observed in the recording, representing a wave traveling the 1200 m between the source and the furthest geophone, in 400 ms. This is a compressional wave propagating along the ground surface at a velocity of 3000 m/s. Two or three lower frequency arrivals are then distinguished, also in a straight line, traveling the 1200 m in time intervals ranging between 700 and 1400 ms. These are surface waves (probably Rayleigh or pseudo-Rayleigh waves), with velocities between 850 and 1700 m/s. A few seismic arrivals aligned with hyperbolas are also observed, particularly an energetic arrival at around 1 s. Their pseudo-period is about 30 ms. These are reflections from the subsurface geological beds. For example, the energetic reflection at 1 s is received from an interface at a depth of 1470 m. The hyperbolic shape of the seismic reflections in the representation (time, distance) is a consequence of the distance of the source/detector spread. We shall examine the types of reflection obtained for a number of specific reflectors.

4.1.6.1 Horizontal reflector in a homogeneous medium

The interpretation of the seismic recordings, such as those in Fig. 4.23, helps to calculate the depth of the reflectors and the propagation velocity of the waves in the subsurface. For a horizontal reflector in homogeneous medium, the arrival times of the seismic reflections are shown as a function of distance in Fig. 4.24. If the source is at S, its image at S', and the detectors at R at distance x from the source, the two-way traveltime of the reflected wave is written:

$$V^2 t^2 = x^2 + 4h^2 \tag{4.1}$$

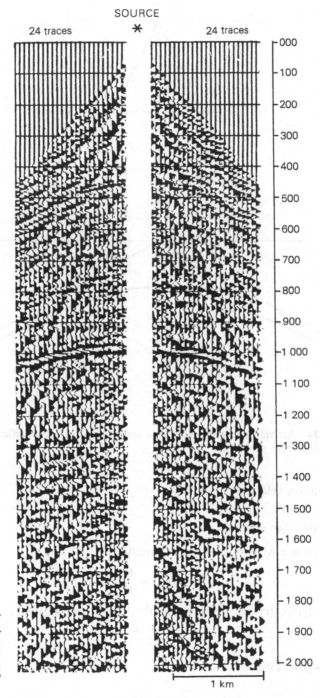

Fig. 4.23 Seismic recording with source at center and 48 traces spread 2400 m on either side of source.

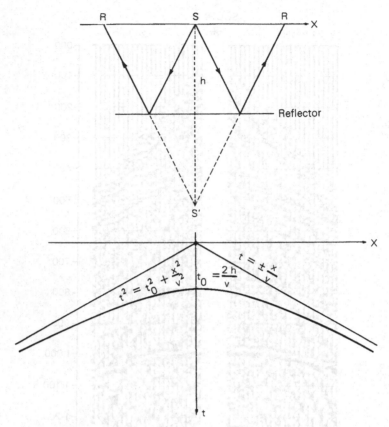

Fig. 4.24 Arrival time of seismic events assuming a horizontal reflector in a homogeneous medium.

where

V = propagation velocity,

h = reflector depth,

x = source/detector distance.

The curve $t(x)$ is a hyperbola with the equation:

$$t^2 = \frac{x^2}{V^2} + \frac{4h^2}{V^2}$$

(4.2)

with an axis of symmetry SS' which allows the lines for asymptotes:

$$t = \pm \frac{x}{V}$$

(4.3)

and for the apex the point:

$$x_0 = 0$$

$$t_0 = \frac{2h}{V}$$

Note that the asymptotes $t = \pm x/V$ represent the direct waves propagating at velocity V from the source to the geophones along the ground surface.

Calculation of propagation velocity and reflector depth

To calculate the propagation velocities and the reflector depths, the first interpreters generally made a representation in the form $t^2(x^2)$ of the seismic arrivals picked from the recordings. They obtained straight lines with slopes $1/V^2$ and ordinates at the origin:

$$t_0^2 = \frac{4h^2}{V^2} \tag{4.4}$$

which gave them an average propagation velocity V (assumed constant) between the surface and the reflector, allowing them to calculate the reflector depth. These calculations are now performed automatically by applying the computer techniques discussed in Section 4.2.7.

4.1.6.2 Dipping reflector in a homogeneous medium

If the reflector is inclined with a dip α, since the propagation velocity V is assumed constant, the two-way traveltime of the reflected wave is given by the equation:

$$V^2 t^2 = x^2 + 4h^2 + 4hx \sin \alpha \tag{4.5}$$

where h denotes the depth of the reflector measured perpendicularly from S (Fig. 4.25).

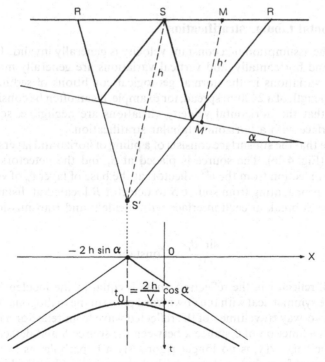

Fig. 4.25 Arrival time of seismic events assuming a dipping reflector in a homogeneous medium.

To find an expression similar to (4.1) which serves to determine the velocity by a representation of the form $t^2(x^2)$, it suffices to express the reflector depth not as a function of h, but as a function of h', the depth measured perpendicularly from the midpoint M of $SR = x$.

Knowing that:

$$h' = h + \frac{x}{2} \sin \alpha \tag{4.6}$$

the two-way traveltime of the reflected wave is written:

$$V^2 t^2 = x^2 \cos^2 \alpha + 4h'^2 \tag{4.7}$$

and the curve $t(x)$ is the hyperbola with equation:

$$t^2 = \frac{x^2}{V_a^2} + \frac{4h'^2}{V^2} \tag{4.8}$$

with

$$V_a = \frac{V}{\cos \alpha} \tag{4.9}$$

The propagation velocity V_a given by the representation $t^2(x^2)$ is slightly higher than the true velocity, with an error up to 3% for a dip of 14°.

4.1.6.3 Horizontal tabular stratification

In practice, the assumption of a constant velocity is generally invalid. The formations vary vertically and horizontally, and vertical variations are generally more substantial than horizontal variations in the normal geological conditions of sedimentary basins. Along the entire length of a 2400 m spread, for example, it can often be considered as a first approximation that the horizontal velocity variations are negligible, so that one can assume a subsurface with a horizontal tabular stratification.

Let us assume that the subsurface consists of a piling of horizontal layers of thickness e_i and velocity V_i (Fig. 4.26). The source is placed at S, and the detectors at R. We are interested in the reflection from the nth reflector, at the base of layer e_n of velocity V_n. The seismic raypath, propagating from source S to detector R located at distance x from the source, undergoes a break at each interface with incident and transmission angles such that:

$$\frac{\sin \theta_i}{V_i} = \text{Constant} \tag{4.10}$$

At point I_n of reflector n, the reflection angle is equal to the incident angle, and the upward paths are symmetrical with the downward paths in the orthogonal symmetry with axis $I_0 I_n$. If the two-way traveltime t of the reflected wave from reflector n is measured on the recordings as a function of distance x between the source S and detector R, it can be observed that the curve $t(x)$ is no longer rigorously a hyperbola, as in the case of the homogeneous subsurface, but a curve with a hyperbolic shape which is called a time/distance curve. Making a representation $t^2(x^2)$ does not rigorously produce a

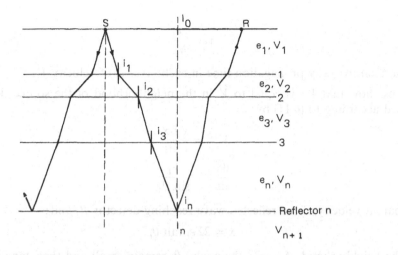

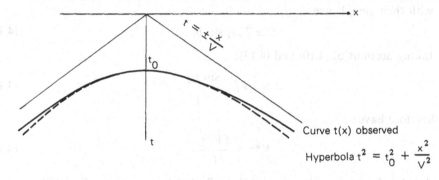

Fig. 4.26 Arrival time of seismic events assuming a piling of horizontal layers.

straight line as in the case of the homogeneous subsurface, but a curve fairly close to a straight line, especially in the neighborhood of the origin $x = 0$. In the neighborhood of $x = 0$, the time/distance curve $t(x)$ approaches a hyperbola with equation:

$$t^2 = t_0^2 + \frac{x^2}{V^{*2}} \tag{4.11}$$

where t_0 is the vertical two-way propagation time from the source to reflector n and V^* is an average propagation velocity. We shall show that V^* is very close to the RMS velocity of the piling of layers.

Root Mean Square velocity

The RMS velocity $\overline{V}$ of a piling of layers of velocities V_i is by definition:

$$\overline{V} = \sqrt{\frac{\Sigma V_i^2 \tau_i}{\Sigma \tau_i}} \tag{4.12}$$

where

$$\tau_i = \frac{2e_i}{V_i} \tag{4.13}$$

is the vertical two-way propagation time in each layer e_i of velocity V_i.

Let us show that V^* is equal to $\overline{V}$ in the neighborhood of the source. V^* can be expressed according to (4.11) by:

$$V^{*2} = \frac{x}{t} \frac{dx}{dt} \tag{4.14}$$

$$\frac{dx}{dt} = \frac{V_1}{\sin \theta_1} \tag{4.15}$$

the apparent velocity of the reflected wave reaching detector R, and:

$$x = 2\Sigma e_i \tan \theta_i \tag{4.16}$$

In the neighborhood of $x = 0$, the angles θ_i remain small and their tangent can be merged with their sine, hence:

$$x \simeq 2\Sigma e_i \sin \theta_i \tag{4.17}$$

and, by taking account of (4.10) and (4.13):

$$x \simeq \Sigma V_i^2 \tau_i \frac{\sin \theta_1}{V_1} \tag{4.18}$$

We therefore have:

$$V^{*2} \simeq \frac{\Sigma V_i^2 \tau_i}{t} \tag{4.19}$$

where t denotes the propagation time of the reflected wave between S and R.

In the neighborhood of $x = 0$, $t = t_0 = \Sigma\tau_i$, the vertical two-way propagation time from the source to reflector n. Hence:

$$V^{*2} \simeq \frac{\Sigma V_i^2 \tau_i}{\Sigma \tau_i} \simeq \overline{V}^2 \tag{4.20}$$

For small source/detector offsets, the time/distance curve $t(x)$ obtained in reflection survey on reflector n is very close to a hyperbola with equation:

$$t^2 = t_0^2 + \frac{x^2}{V_{(n)}^2} \tag{4.21}$$

where $\overline{V}$ denotes the RMS velocity of the piling of layers.

An RMS velocity $V(n)$ exists for each of the reflectors n. It is written:

$$\overline{V}(n) = \sqrt{\frac{\displaystyle\sum_{i=1}^{n} V_i^2 \tau_i}{\displaystyle\sum_{i=1}^{n} \tau_i}} \tag{4.22}$$

Remark: The RMS velocity $\overline{V}(n)$ must not be confused with the average velocity:

$$V_M(n) = \frac{\sum\limits_{i=1}^{n} V_i \tau_i}{\sum\limits_{i=1}^{n} \tau_i} \tag{4.23}$$

In the stratified formations normally encountered in nature, the difference between the average velocity V_M and the RMS velocity $\overline{V}$ is generally a few per cent. It can be shown (Cordier, 1983) that $\overline{V}$ is always slightly higher than V_M. The RMS velocities allow the "interval" velocities of each of the layers of the piling to be obtained, using the Dix equation:

$$V_n^2 = (\overline{V}_{(n)}^2 t_n - \overline{V}_{(n-1)}^2 t_{n-1})/(t_n - t_{n-1}) \tag{4.24}$$

4.1.7 Three-dimensional surveys

Two-dimensional surveys can only yield strictly accurate results if the subsurface structures are two-dimensional, (monoclines, elongated anticlines and synclines, faults and flexures with a vertical plane of symmetry.) To obtain detailed information about structures with no vertical plane of symmetry, three-dimensional surveys are necessary.

On land, the standard implementation of three-dimensional surveying consists in spreading the source lines perpendicular to the detector lines (Fig. 4.27). Several parallel lines of detectors can be used simultaneously to obtain multiple CMP coverage over rectangular areas. A more elaborate method consists in laying out sources and detectors uniformly around a square. With an equal number of sources and detectors, a zone of common midpoints is obtained covering an entire square. In the case shown in Fig. 4.28 a, for example, for 8 source points and 8 detection points merged and distributed uniformly over the perimeter of the square, 25 reflector points are theoretically obtained covering the entire area of the square. In practice, square grids of 24 sources and 24 detectors can be used, distributed on the perimeter of a 1200 m sided square (Fig. 4.28 b).

Another example of this method consists in using source points uniformly spaced on either side of a grid of detectors (Fig. 4.29). This technique helps to perform 3D surveying with a high degree of CMP multiplicity at relatively low cost. In areas where only the roads and paths are accessible, the spreads are less regular and the degree of CMP coverage varies considerably. The implementation of 3D land surveying may therefore give rise to a wide variety of spreads. At all events, the position of the sources and detectors must be known perfectly in order to associate a common midpoint with each source/detector pair, so as the relocate the different reflectors in space.

At sea, the most widely-used method consists in recording a series of parallel profiles at 100 to 200 m intervals with a single ship. Marine currents are often sufficient to impart to a 2400 m streamer lateral drifts of more than 100 m at the streamer tail, giving three-dimensional CMP coverage in a relatively narrow volume vertically below the streamer. The streamer must be equipped with compass systems capable of measuring its drift accurately. A shipboard computer calculates the position of all the hydrophones at any time, and consequently the position of all the common midpoints with respect to the ship (Fig. 4.30). These common midpoints must be distributed as uniformly as possible to

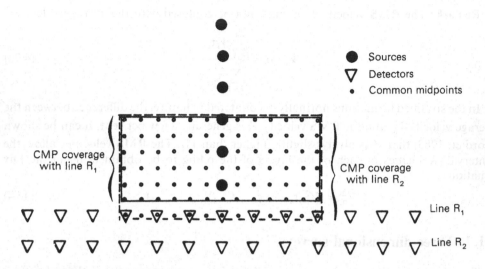

Fig. 4.27 Three-dimensional seismic surveying onshore along perpendicular source and detector lines.

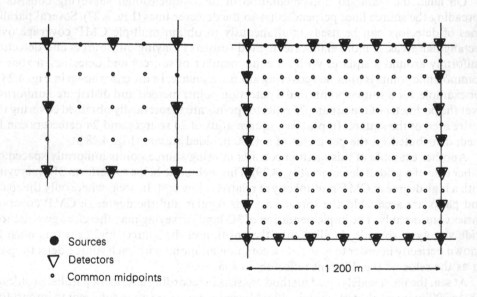

(a) First example of application by square spread for eight source and detection points distributed on the perimeter of the square.

(b) Second example of application by square spread for twenty-four source and detection points distributed on the perimeter of the square.

Fig. 4.28 Three-dimensional seismic surveying onshore, along square grid.

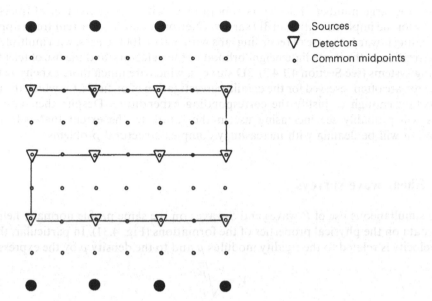

Fig. 4.29   Three-dimensional seismic surveying onshore with source points outside the detector grid.

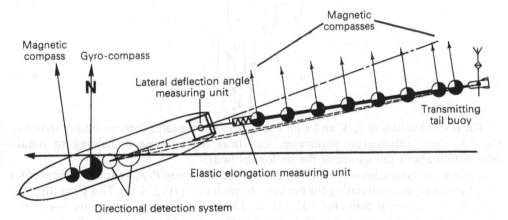

Fig. 4.30   Three-dimensional seismic surveying offshore with use of streamer drift. (*Prakla-Seismos* Document).

avoid the risk of having zones without seismic raypaths. The determination of all the common midpoints, or binning, generally takes place in real time (namely between two successive emissions), making it possible to check that every part of the subsurface has been scanned by the seismic raypaths.

Another technique sometimes used in 3D marine surveys consists in using several sources, laterally offset on both sides of the ship, which emit seismic signals alternately. This provides three-dimensional data even in the absence of streamer drift.

Given the large number of detectors, laboratories with a large number of traces are necessary for the implementation of 3D surveys (Nelson, 1983). The current trend appears to be heading towards the use of recording labs with 500 or 1000 traces, with multiplexing and digitization in the immediate neighborhood of the detectors, and the use of telemetry recording systems (see Section 4.1.4.3). 3D surveys, which are much more expensive than 2D surveys, are often reserved for the detailed investigation of oil fields, for which the issue is important enough to justify the corresponding expenditure. Despite their cost, 3D surveys will probably see increasing use in the future to the extent that petroleum exploration will be dealing with increasingly complex structural problems.

4.1.8 Shear wave surveys

The simultaneous use of P-waves and S-waves on the same profile normally helps to obtain data on the physical properties of the formations (Fig. 4.31). In particular, the S-wave velocity is related to the rigidity modulus μ and to the density ρ by the expression:

$$V_s = \sqrt{\frac{\mu}{\rho}}$$

and the ratio V_p/V_s is related to Poisson's ratio σ by the formulas:

$$\frac{V_p}{V_s} = \left(\frac{1-\sigma}{\frac{1}{2}-\sigma}\right)^{\frac{1}{2}} \qquad (4.25)$$

$$\sigma = \frac{1 - \frac{1}{2}\left[\frac{V_p}{V_s}\right]^2}{1 - \left[\frac{V_p}{V_s}\right]^2} \qquad (4.26)$$

The normal values of V_p/V_s and σ for commonly encountered formations are shown in Fig. 4.32. New lithological parameters can thus be determined, helping to obtain information about the nature of the geological beds.

Shear wave seismic surveys generally consist in conducting P-wave and S-wave profiles simultaneously, and correlating the horizons to each other (Fig. 4.33). This gives the ratio V_p/V_s of the geological beds (Polchkov et. al., 1980). The S-wave sources are horizontal vibrators, which transmit a horizontal motion to the ground by means of a vibration plate, or impactors, like the IFP "Marthor", which transmit the horizontal motion to the ground by a hammer horizontally striking a baseplate anchored to the ground.

a. SH-waves

SH-waves, in which the particle motion is polarized perpendicular to the vertical plane of the profile, are normally generated by directing the source perpendicular to the profile. If the stratification is tabular, or such that the vertical plane of the profile is a plane of symmetry for the layers, no conversion into P-waves occurs at the interfaces, and the SH-waves are theoretically reflected and transmitted exclusively as SH-waves. The detectors are generally horizontal geophones, turned perpendicular to the profile (Fig. 4.34 b).

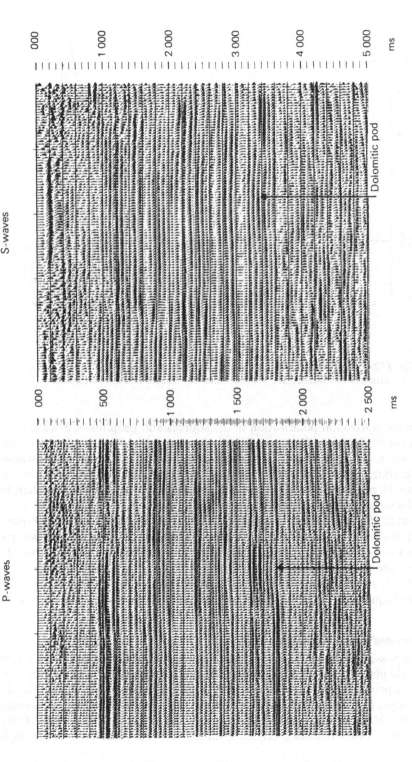

Fig. 4.31 *P*-wave and *S*-wave seismic profiles recorded at the same location. (*CGG Document*).

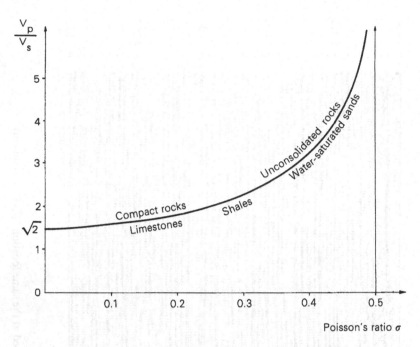

Fig. 4.32 V_p/V_s Ratio versus Poisson's ratio σ and zones where these parameters are located for some commonly encountered formations.

Apart from the horizontal polarization of the sources and geophones, S-wave shooting spreads are similar to those of P-waves. Note however that the S-wave velocities are lower than the P-wave velocities (about 1/2 in compact sediments, 1/3 to 1/4 in unconsolidated surface formations), and the surface waves are often more significant (Love waves). The wavenumber filtering patterns are hence often more elaborate for S-waves than for P-waves, and source/geophone offsets are higher.

The emitted frequency band is generally lower with S-waves than with P-waves to preserve equivalent wavelengths. With horizontal vibrators, for example, the sweep often extends from 8 to 40 Hz for S-waves, compared with sweeps of 12 to 60 Hz or 16 to 80 Hz for P-waves. The resolution of S-wave surveys is of the same order of magnitude as that of P-wave surveys, and the reduction of propagation velocities is often close to the reduction of the useful frequencies. However, the penetration of S-waves is often less than that of P-waves.

b. ***SV-waves and converted waves***

It is also possible to carry out SV-wave and converted wave surveys. The horizontal geophones are placed parallel to the profile, and the source is either a horizontal vibrator parallel to the profile for SV-waves, or a vertical vibrator for converted waves (Fig. 4.34 c). In the latter case, the waves are emitted vertically for P-waves, and undergo conversions to SV-waves, by reflecting from the subsurface interfaces at oblique incidences. Once the source/geophone offset reaches a certain distance (in practice as soon as the angles of

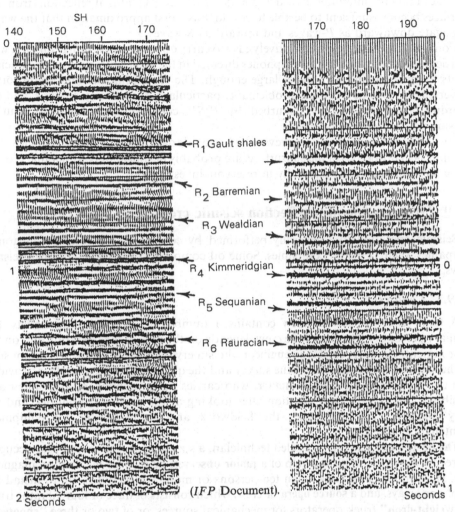

Fig. 4.33   *P*- and *SH*-wave profiles, and calibration at top of geological horizons.

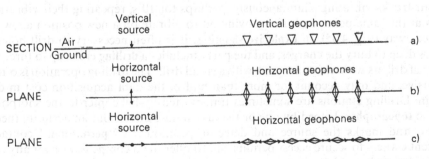

Fig. 4.34   *P*-, *SH*- and *SV*-wave spreads.   (a) *P*-waves; (b) *SH*-waves, and (c) *SV*-waves.

incidence upon the interfaces exceed about 15°), P/SV conversions at reflections from the interfaces become sufficient to be able to consider as a first approximation that the waves propagate downward as P-waves and upward as S-waves.

Converted wave surveys are relatively easy to carry out, simply requiring emission with a vertical source and horizontal geophones directed in the profile direction, with sufficient offsets to make the incident angles large enough. The processing and interpretration of converted waves is nevertheless problematic, particularly the velocity analysis, and the recordings are liable to be disturbed by P/SV conversions which occur during transmission accross the interfaces.

S-wave and converted wave surveys, which provide data about certain lithological and petrophysical properties of the subsurface, are probably destined for considerable growth in the forthcoming years, especially in reservoir investigation.

4.1.9 Organization of a reflection seismic crew

Seismic operations are generally performed by contractors on behalf of customer companies, such as the oil companies. Some oil companies even have their own seismic crews.

4.1.9.1 Land crew

A land reflection seismic crew contains a number of geophysical specialists and auxiliary personnel. The size of the crew depends on the area of the operation and on the environment. The **party chief** is an engineer who supervises the staff and who is responsible for the satisfactory completion of the survey and the quality of the results. He is provided with two assistants, the **chief computer,** who carries out a preliminary assessment and possibly a preliminary interpretation after looking at the playback monitors, and the **party manager,** who supervises the fieldwork and is responsible for equipment maintenance and safety.

The **observer** is generally a skilled technician, a specialist in electronics, who occupies the recording truck. With the help of a **junior observer,** he records the data on magnetic tape. He heads a **cable crew,** often ten persons or more, whose duties are to install the geophone arrays, and a **source operator crew,** which consists of three or four vibrator truck or "**weight-drop**" truck operators for mechanical sources, or of two or three "**shooters**" for explosive sources.

In **vibroseismic** prospecting, it is customary to see along the roads and in the fields three or four trucks vibrating simultaneously, perhaps for 10 s, repeating their vibrations six times at the same point, and then moving on to vibrate on a new position a few meters further away (Fig. 4.35). In **explosive shooting,** it is often necessary to drill holes a few meters deep to bury the charges, and the party includes a **drilling crew,** up to three or four teams of drillers with assistants, each with a small drill. The drilling operation is often very expensive, and may account for more than half of the data acquisition cost in difficult terrain. Drilling stations are sometimes transported by helicopter to the shotpoints.

The **topographer,** helped by one or two assistants, carries out an accurate theodolite survey, and marks the source and detector positions. A "**permitman**" contacts the different owners to secure access permits and to offer them compensation for any damage caused. A number of **laborers** prepare the ground, transport equipment (the transport of

Fig. 4.35 "Vibroseis" survey: three vibrators in line. (*Sercel* Document).

water in desert areas is a heavy burden), perform camp maintenance, etc. Many workers may be needed, more than one hundred in hostile environments.

Liaison between the party and the client company is performed by a **supervisor,** an engineer of the client company, often present in the field, who checks permanently the quality and effectiveness of the operations. A land crew may have as many as ten geophysicists, together with auxiliary personnel often recruited locally.

4.1.9.2 Marine crew

The marine seismic crew is generally accommodated onboard a medium-sized ship (about 1000 t) which tows the sources and detectors and houses the recording lab (Fig. 4.36). The geophysical crew includes:

(a) The **party chief,** who is responsible for the success of the survey and for the quality of the work.

(b) The **observer,** assisted by several **junior observers,** who carry out permanent data acquisition, day and night, relaying each other by shifts.

(c) Several **technicians** who perform maintenance and immersion of the sources and the streamer, operations that are often difficult, particularly in heavy seas.

(d) A total of about ten geophysicists are aboard, plus the client company representative, who is often a geologist-geophysicist familiar with the area. Added to these geophysicists is a comparable number of seamen.

Navigation is often automatic, run by the shipboard computer, which integrates the radio navigation, Doppler sonar and gyro-compass data with the satellite results. The "shot" sequence is programmed to fire at predetermined positions, every 12.5 m for example, along the profile. Some teams have specialized computers designed to preprocess the data in real time, onboard. This allows an effective check of the quality of the results, optimization of the implementation schedule, a preliminary interpretation onboard, and the preparation of the data processing to be carried out on the central computer onshore. The ship is under the responsibility of a captain, who follows the instructions of the party chief. The captain is always responsible for safety.

Fig. 4.36 Geophysical prospecting vessel *Résolution* (*IFP* Document).

4.1.10 Implementation cost

A reflection seismic survey represents a considerable mobilization of personnel and equipment, several tens or even hundreds of persons, and some twenty vehicles onshore, and a ship displacing about 1000 t offshore. The cost of data acquisition is hence very high, around $500,000 a month, onshore and offshore (1986 figures). Since the distances covered are usually much higher offshore, 100 km/day as opposed to a few kilometers onshore, the cost of acquisition in 1986 amounted to between $3000 and 8000/km onshore, and between $100 and $300/km offshore. Note that the processing cost is also higher onshore than offshore due to the number of corrections required in surface zones on land. This processing, which costs around $1500 to $3000/km onshore, costs only $300 to $500/km offshore.

On the whole, the total cost per kilometer of the survey, including data acquisition and processing, is much higher onshore, $5000 to $10,000/km, as compared with $500 to $1000/km offshore. If this is compared with the cost of petroleum drilling, which is generally very high offshore, the cost of marine seismic surveys is sufficiently low to justify its systematic implementation for the exploration and development of offshore fields. Even three-dimensional surveys, which we have seen to be more expensive due to the density of the profiles and the complexity of the processing, is justified if it helps to economize one or more drilling operations offshore.

4.2 PROCESSING

Processing is applied to the seismic data recorded in the field in order to highlight the geological features. The magnetic tapes from the field are sent to the data processing center, where they are demultiplexed and placed in a format compatible with the computer. The normal processing sequence comprises frequency filtering, amplitude recovery, deconvolution before stack, static corrections, CMP gathering, velocity analysis, moveout correction, CMP stacking, deconvolution after stack, and eventually migration. Furthermore, additional operations can be performed, such as cross-correlations with sweep signal in vibroseismic prospecting. After processing, the seismic traces are played back and assembled in the form of variable-area sections, for a clear representation of the subsurface geological features.

4.2.1 Demultiplexing

Demultiplexing consists in separating the individual seismic channels, reassembling the series of samples corresponding to each detector sequentially. The result of demultiplexing is often recorded on a new magnetic tape called the demultiplexed tape, on which are also entered data such as the profile number, shot number, trace number, etc. These are the identification "headers". The time sampling interval is usually 4 ms in a conventional survey, and 1 or 2 ms in a high resolution survey.

4.2.2 Sweep signal correlation

In vibroseismic prospecting, the signal is a long frequency-sweep wave train, and it must be compressed by cross-correlating the received signal with the emitted sweep signal (see Section 3.6.2.2). This operation is performed by the computer, using the entire dynamic range of the recording (12 mantissa bits + 4 gain bits, for example), to achieve maximum enhancement of the signal-to-noise ratio. In order to verify that the recording operation is correct, an approximate cross-correlation is made in the field, with a limited number of bits (e.g. 12 bits), immediately yielding a "monitor" similar to those obtained in impulse surveys (Fig. 4.37). Certain correlators now allow field correlation in real time with maximum accuracy.

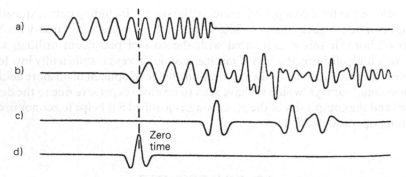

Fig. 4.37 Example of Vibroseis signal cross-correlation. (From Anstey, 1981, *Seismic Prospecting and Instrument Specification*, Vol. 1, Gebrüder Bornstraeger, Berlin).
(a) Sweep signal; (b) Reflected signal before cross-correlation; (c) Reflected signal after cross-correlation with sweep signal and (d) Autocorrelation of sweep signal.

4.2.3 Gain recovery and attenuation correction

For each sample, the knowledge of the level of the recorded signal and its gain helps to recover the true amplitude of the signals reaching the detectors. The signals become weaker with increasing reflector depth. Attenuation correction of the deep reflections is often carried out by empirical formulas, or by automatic gain controls which adjust the signal amplitudes in windows of 200 or 300 ms duration to identical average values (Fig. 4.38).

An attempt is sometimes made to compensate separately for the attenuation due to **geometric spreading,** namely the decrease in amplitude due to the energy distribution on the expanding wavefronts, and the other causes of attenuation. If the medium were homogeneous with constant velocity and without absorption, the amplitude would be proportional to $1/r$, where r is the distance traveled by the wave. In practice, the velocity

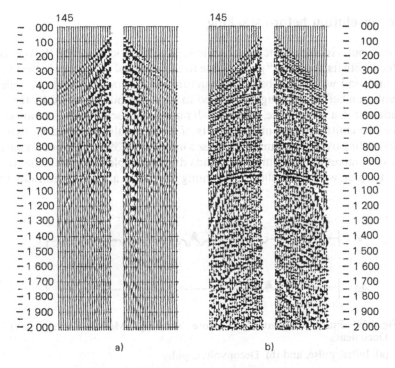

Fig. 4.38   Record before and after attenuation correction. (*IFP* Document).
(a) Raw recording, source at center, and (b) After low-cut filtering and
attenuation correction by automatic gain control.

increases with depth, and the amplitude falls more rapidly. We showed in Section 2.6 that
geometric spreading could be expressed by a term $V_1/t\overline{V}^2$ as a function of the two-way
propagation time t, the RMS velocity $\overline{V}$ between the surface and the reflector, and the
velocity V_1 in the surface formations (Newman, 1973). It is difficult to use this formula at
this stage of processing, since the velocity analysis that serves to determine the RMS
velocities has not yet been performed. In difficult cases, where the amplitudes must be as
representative as possible of the reflectivity of the reflectors (to calculate the acoustic
impedances, for example), a preliminary approximate processing is necessary, with
attenuation correction by automatic gain control and approximate determination of the
RMS velocities, followed by a more precise amplitude correction with the Newman
formula or a similar formula.

One can also try to compensate for the attenuation due to **absorption,** namely the
attenuation independent of geometric spreading of the wavefronts. This often ranges
between 0.1 and 0.5 dB per wavelength traveled in the ground, and depends on the seismic
signal frequency. A gain function whose spectrum varies with time must be applied to the
seismic signal, strengthening the high frequencies even more with increasing reflector
depth. Here also, this correction is applied only in difficult cases, when data of a
stratigraphic and lithological character are desired.

4.2.4 Deconvolution before stacking

The seismic pulse is often stretched by various factors associated with the type of source (bubble effect), with its position ("ghosts" due to the image of the source in the free surface of the ground), and with the subsurface structure (reverberations, multiple reflections).

Deconvolution before stacking is designed to spike the pulse emitted by the source in order to reduce it to a short pulse, with a small number of oscillations. In marine surveys, for example, the emitted signal often consists of a short pulse followed by two or three rebounds extending over a few hundred milliseconds (Fig. 4.39 a). This occurs particularly with explosive sources and air guns. Rebounds due to bubble effect make the recordings difficult to interpret, with each reflector assuming the appearance of several superimposed reflectors.

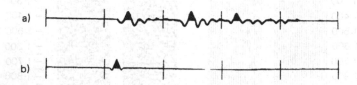

Fig. 4.39 Pulse deconvolution before stacking. Marine survey. (*GSI* Document).

(a) Initial pulse, and (b) Deconvolved pulse.

The deconvolution operator must spike the signal into a short pulse. It is generally calculated by a least squares method, in which:

$$s_1, s_2, \dots s_l = \text{sampled initial signal}$$
$$f_1, f_2, \dots f_n = \text{desired deconvolution operator}$$
$$r_1, r_2, \dots r_m = \text{desired final pulse}$$

The result of the deconvolution is symbolically written:

$$r_i = s_i * f_i \qquad (4.27)$$

where * denotes the convolution symbol.

The least squares method consists in summing up the squares of the differences:

$$\Sigma = \sum_{i=1}^{m} [r_i - (s_i * f_i)]^2 \qquad (4.28)$$

between the desired pulse r_i and the pulse obtained $s_i * f_i$, and in minimizing it by canceling its partial derivatives with respect to the variables f_i. This yields n equations of the form:

$$\frac{\partial}{\partial f_j} \sum_{i=1}^{m} [r_i - (s_i * f_i)]^2 = 0 \qquad (4.29)$$

for $j = 1$ to n, which serve to determine the n unknowns, f_1, f_2, ... f_n of the deconvolution operator. The seismic traces y_i are then convolved by the operated f_i and the deconvolved traces z_i are obtained:

$$z_i = y_i * f_i \qquad (4.30)$$

This method presumes that the signal s_i delivered by the source is known. It can sometimes be measured, especially in marine surveys, by placing hydrophones near the source. It can also be estimated, by computing the autocorrelation of the seismic traces, which often closely approaches the autocorrelation of the signal emitted by the source. Knowing an approximation of the autocorrelation of the emitted signal, and making an assumption on its phase, one can compute the emitted signal. It is often necessary to compute a deconvolution operator at each shot, to take into account the pulse variations due to source depth fluctuations.

4.2.5 Static corrections

Mainly necessary in land surveying, static corrections consist in correcting the traveltime anomalies introduced by the variations in geophone altitude and by the velocity variations in the surface formations.

4.2.5.1 Field statics

These consist in adjusting the traveltime to what it would be if the sources and geophones were located on an arbitrary reference plane (datum plane) slightly below the weathered zone. The values of the corrections are easily computed if the elevation of the sources and geophones, the propagation velocity, and the thickness of the weathered zone are known. Two simultaneous methods are generally employed:

(a) **Seismic shots** are fired at the base of the weathered zone, at a depth of 10 to 15 m, and a surface geophone gives an accurate measurement of the traveltime in this zone. This is a very accurate method, but can be expensive in the case of many lateral velocity variations.

(b) **Small refraction seismic profiles** are implemented to investigate the weathered zone (see Section 6.1.1). The geophone layouts employed are spread over about 100 m. Direct shots and reverse shots are fired with small explosive charges, a few tens of grams or so, or weights are dropped, with 200 kg dropping from a height of 2 m on a target coupled with the ground, for example. This method gives the propagation velocity and the thickness of the weathered zone.

Using (a) and (b) simultaneously serves to compute the static corrections with relatively good accuracy.

4.2.5.2 Residual statics

Once the "field static" corrections are made, "residual" traveltime statics often subsist due to localized velocity variations below the geophones and below the sources. These anomalies, often slight (a few milliseconds), must be corrected to obtain homogeneous, consistent recordings.

These corrections are called "residual statics". By comparing the seismic traces obtained from the same shotpoint and different detection points, after applying the normal moveout corrections (see Section 4.2.8), shifts are observed in the reflection arrival times, due essentially to traveltime anomalies in the neighborhood of each geophone. The residual statics at the geophones are computed by correcting these shifts, after having measured them by standard cross-correlation procedures (Fig. 4.40). The residual statics at the shotpoints are computed in the same way by comparing the seismic traces obtained from the same detection point and from different shotpoints.

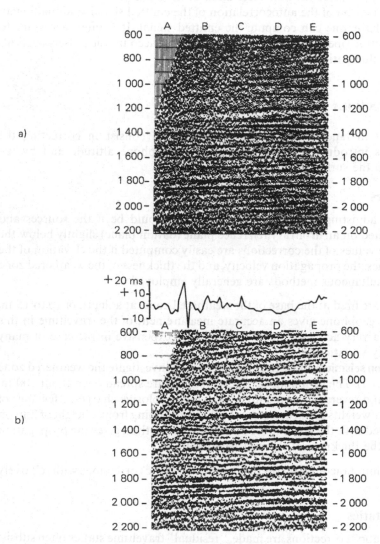

Fig. 4.40   Static corrections. (*CGG* Document).

(a) Raw section, and (b) Section after field + residual static corrections. The corrections applied are shown at the top of plate (b).

4.2.6 Common midpoint gathering

Up to this stage of processing, the seismic traces are grouped in collections of shotpoints, 48 or 96 seismic traces per shotpoint, for example (Fig. 4.3). CMP gathering consists in rearranging the traces in a different order so as to group all those that have common reflection points (Fig. 4.41). The sort, carried out by the computer according to the identification headers, consists in finding trace R_1 of shot S_1, trace R_2 of shot S_2, trace R_3 of shot S_3, etc., and storing them sequentially in a new file.

Note that no common reflection point exists if the reflectors are sloping, because each $S_i R_i$ pair gives rise to a different reflecting point (Fig. 4.42). This is why it is preferable to speak of common midpoints.

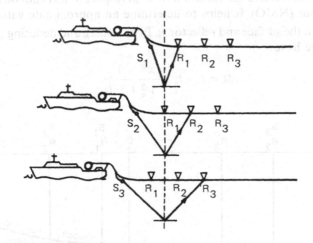

Fig. 4.41   CMP collection. Marine survey.

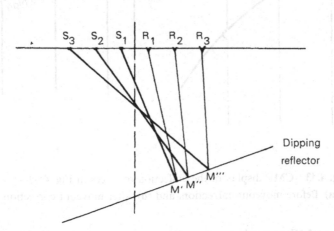

Fig. 4.42   CMP collection and positions of reflecting points for a dipping reflector.

4.2.7 Velocity analysis

In formations with horizontal stratification, the $t^2(x^2)$ analysis described in Section 4.1.6.3 serves to determine the RMS velocities, irrespective of the trace array, common shotpoint, common detection point, or common midpoint. In randomly stratified formations, where the beds may have varying dips, it can be shown (Levin, 1971) that only CMP trace arrays can be used to obtain the RMS velocities with sufficient accuracy. The error does not exceed 3% for dips of around 15°.

Let us represent the CMP traces as in Fig. 4.43, where the distance SR is plotted on the abscissa and the two-way propagation time t on the ordinate. The abscissa point $x = SR$ is shown symbolically by SR. A hyperbolic curve is obtained for each reflector. Curve $t(x)$ is the time/distance curve, and the variation Δt with respect to its ordinate at the origin is the normal moveout (NMO). It helps to determine an approximate value of the RMS velocity $\overline{V}_n$ between the surface and reflector n. From (4.21), and denoting the ordinate at the origin by t_n, we have:

$$\Delta t = t - t_n \cong \frac{x^2}{\overline{V}_n^2} \frac{1}{t + t_n} \tag{4.31}$$

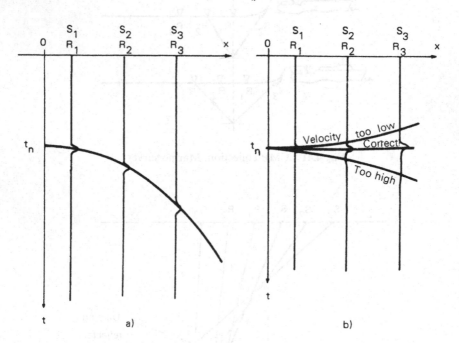

Fig. 4.43 CMP display of the collection of traces in Fig. 4.42.
(a) Before moveout correction, and (b) After moveout correction.

Δt is called the **NMO correction** because it is the correction that must be applied to t to make the time/distance curve horizontal. For an RMS velocity $\overline{V}_n = 2000$ m/s, for

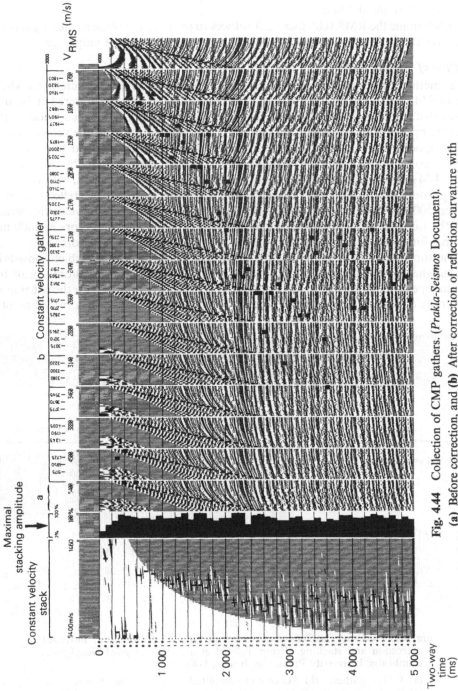

Fig. 4.44 Collection of CMP gathers. (*Prakla-Seismos* Document).

(a) Before correction, and **(b)** After correction of reflection curvature with different velocities.

example, a reflection at $t = 1$ s and a source/geophone distance $x = 1$ km, the NMO correction Δt is about 120 ms.

To determine the RMS velocities V_n, it suffices to measure the slope at the origin of all the curves $t^2(x^2)$. In practice, much faster computer techniques are employed.

Principle of velocity analysis

The method consists in carrying out the NMO correction on the computer for a whole series of RMS velocities, from the slowest to the fastest that can be encountered in the area investigated. Figure 4.43 b shows that, if the corrected velocity is the true velocity, the reflector appears to be horizontal after normal moveout correction. Il the velocity is too high, it appears inclined downward, and if too low, it appears inclined upward. It is therefore possible to estimate the RMS velocity of each reflector visually by this method. Figure 4.44 shows a collection of CMPs before and after NMO correction, with velocities ranging from 1450 to 5400 m/s. The reflectors are located at two-way reflection times between 500 and 3500 ms, and it may be observed that the corrected velocities which make the time/distance curves horizontal are about 1900 m/s for reflectors at 1000 ms, 2660 m/s for reflectors at 2200 ms, and 3140 m/s for the deepest reflectors.

An automatic method can also be used to estimate the RMS velocity, which consists in stacking the corrected traces or measuring their coherence. An amplitude peak (or coherence peak) exists for the correct velocities. The stacking amplitude (or the coherence) is generally computed in windows of about 100 ms. Figure 4.45 shows a recording of a

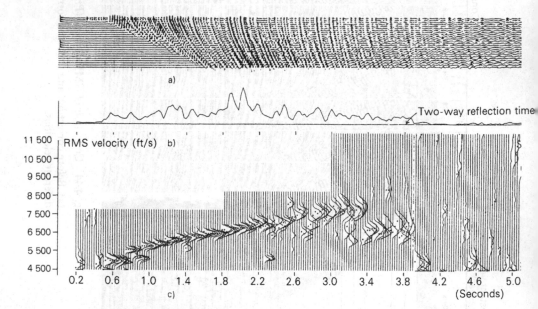

Fig. 4.45 Velocity analysis moveout scan by CMP collection, moveout correction and stacking. (From Telford *et al.*, 1976, *Applied Geophysics*, Cambridge University Press, Cambridge, UK).

(a) CMP gather; (b) Maximum amplitude in stacked traces, and (c) Stacked trace amplitudes as a function of stacking velocity in 100 ms windows.

CMP collection and, every 25 ms, the amplitudes of the stacked traces in 100 ms windows, as a function of the NMO correction velocity.

Figure 4.45 serves to determine the distribution $\overline{V}(t)$ of the RMS velocities as a function of two-way reflection time. Note in this figure that the RMS velocities increase steadily with depth, rising from 4500 to 8000 ft/s as the depth increases from 0.6 to 3.4 s. Between 3.4 and 4 s, much lower velocities appear, around 6500 ft/s. This is probably due to multiple reflections which, although arriving at relatively later times, have propagated in slower formations, explaining their low velocity. This shows that the velocity analysis can be highly effective for identifying multiple reflections.

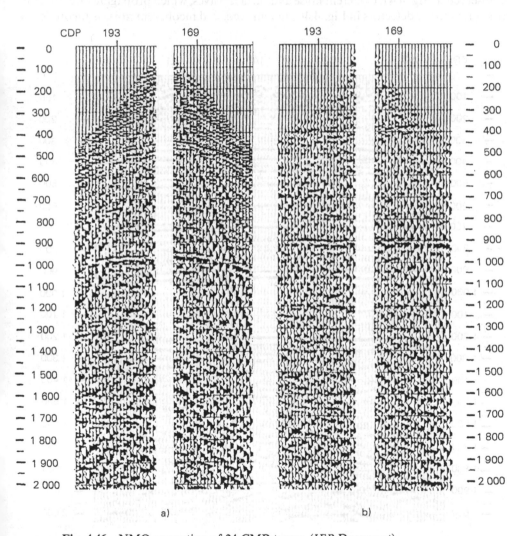

Fig. 4.46 NMO correction of 24 CMP traces. (*IFP* Document).

(**a**) Before correction, and (**b**) After NMO correction. Corrected traces are shifted upward 50 ms.

4.2.8 NMO correction and CMP

Once the distribution $\overline{V}(t)$ of the RMS velocities as a function of reflector depth is determined, NMO corrections can be applied to each of the trace collections and stacking performed. For approximately 24, 48 or 96 coverages, 24, 48, or 96 CMP traces are stacked. Thus the 24 traces in Fig. 4.46 a are corrected in Fig. 4.46 b and stacked in a single trace, trace 175 in Fig. 4.47.

CMP stacking achieves a significant improvement in the signal-to-noise ratio, as may be observed in Fig. 4.47. Coherent noise and surface waves, which propagate directly from the sources to the detectors in Fig. 4.46 are removed, and incoherent noise is theoretically

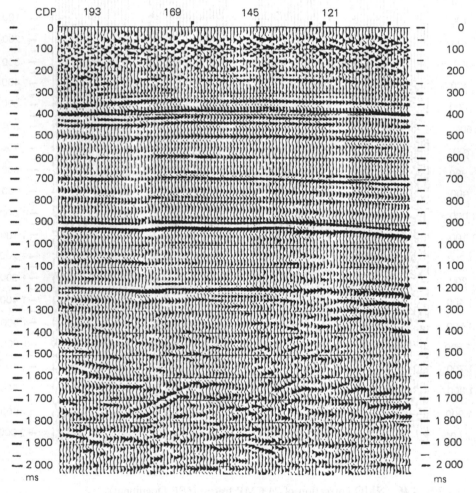

Fig. 4.47 Stacked section 24-fold CMP (50 m between traces). (*IFP* Document).

attenuated in the ratio $\sqrt{n}$, where n is the degree of CMP coverage. Moreover, the multiple reflections are attenuated, as the velocity selected for the NMO correction is the velocity of the primary reflections, and not that of the multiples. The CMP stacked section is the equivalent of the fictive section that would be obtained if the sources and detectors were placed at the CMP, with the seismic raypaths returning on themselves after reflection in normal incidence upon the reflectors (Fig. 4.48).

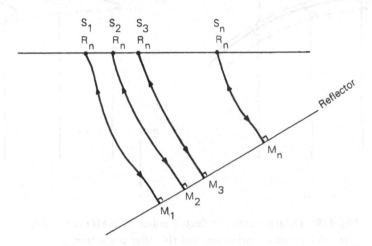

Fig. 4.48   Raypath simulation of CMP stacked traces after correction.

The need for muting

Muting is the zero setting of trace portions before stacking in which the signal is liable to undergo excessive deformations by the NMO correction. Large NMO corrections, like those made for distant traces and shallow horizons, deform the reflected pulses, because the NMO correction:

$$\Delta t = \frac{x^2}{V^2} \frac{1}{t + t_n}$$

is greater for the first arch of the signal than for the second. A 30 ms pulse, for instance, can thus be stretched to 35 ms after NMO correction (Fig. 4.49). It is very difficult to avoid this deformation without excessive complication of processing, and one can simply decide to "mute" the trace portions in which the signal risks excessive deformation. In Fig. 4.49, for example, the trace portion $S_3 R_3$ is set to zero between the origin and 200 ms.

Another advantage of muting is the inhibition of spurious refraction arrivals. In fact, it is on the distant traces and for shallow horizons that refraction arrivals are observed that have propagated along fast reflectors and reach the geophones before shallow reflections. Since the reflection arrivals obtained should be as pure as possible, muting helps to remove the trace portions in which the reflections are disturbed by refractions. As a rule, muting before CMP stacking helps to remove all or part of a trace disturbed by noise. Since seismic recordings with a high order of multiplicity are normally available, 48 to

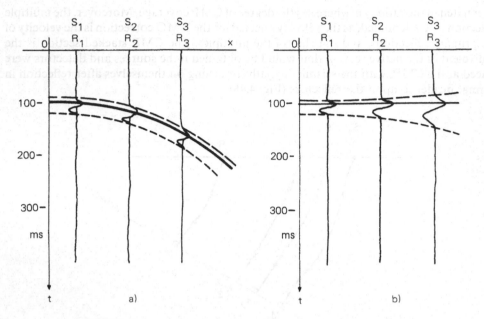

Fig. 4.49 Deformations of reflected pulses by NMO correction.

(a) Pulses before correction, and (b) After correction.

96 CMP coverage for example, a sufficient number of traces always remains to give the reflection information. Suitable weighting is nevertheless necessary to take account of the trace portions removed in order to preserve the true amplitudes of the reflectors after stacking.

4.2.9 Deconvolution after stacking

Deconvolution is often intended to remove secondary reflections that are insufficiently attenuated by CMP stacking, such as reverberations of the water layer in marine surveys.

4.2.9.1 Backus deconvolution

This relatively old method (Backus, 1959) consisted in applying an operator to each trace, computed from the knowledge of the reverberations.

Let:

k = water bottom reflection coefficient,

$\delta(t)$ = seismic pulse, which can be assumed to be a Dirac pulse without limiting the problem,

τ = two-way traveltime in the water layer: τ is constant if constant water depth and normal incidence are assumed.

The signal descending to the reflectors has the form:

$$f(t) = \sum_{n=0}^{\infty} (-1)^n k^n \delta(t - n\tau) \tag{4.32}$$

and its Fourier transform is written:

$$F(\omega) = \sum_{n=0}^{\infty} (-1)^n k^n \exp(-jn\omega\tau) \tag{4.33}$$

$$F(\omega) = \frac{1}{1 + k \exp(-j\omega\tau)} \tag{4.34}$$

where

$f(t)$ = impulse response of the water layer,
$F(\omega)$ = frequency filter formed by this layer.

To **remove reverberations at the source,** the inverse filter $G(\omega)$ must be applied such that:

$$F(\omega) . G(\omega) = 1 \tag{4.35}$$

or

$$G(\omega) = 1 + k \exp(-j\omega\tau) \tag{4.36}$$

Its impulse response is:

$$g(t) = \delta(t) + k\delta(t - \tau) \tag{4.37}$$

Hence the seismic trace must be convolved by the operator $g(t)$ consisting of the two terms 1 and k (Fig. 4.50).

To **remove reverberations at the source and receiver,** the inverse filter $G(\omega)$ must be applied twice. The antimultiple filter at the source and at the receiver is written:

$$G_2(\omega) = (1 + k \exp(-j\omega\tau))^2 \tag{4.38}$$

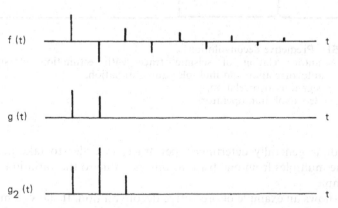

Fig. 4.50   Water layer reverberation and Backus filter:
$f(t)$ = reverberation at source,
$g(t)$ = filter for source only,
$g_2(t)$ = filter for source + reception.

and its impulse response:

$$g_2(t) = \delta(t) + 2k\delta(t - \tau) + k^2\delta(t - 2\tau) \tag{4.39}$$

The seismic trace must therefore be convolved by the operator $g_2(t)$ consisting of the three terms 1, $2k$ and k^2.

4.2.9.2 Predictive deconvolution

There is a growing tendency to use "predictive" methods, which help to determine the appropriate operator to be applied for each trace rapidly. Predictive methods consist in using the information from the early part of the seismic traces to predict the reverberations and multiples, and then to remove them (Peacock and Treitel, 1969). They often use the trace autocorrelation. Assuming for example that the trace autocorrelation can be treated as the autocorrelation $a(t)$ of the signal affected by the multiples, the early part $a_0(t)$ of the autocorrelation is treated as the autocorrelation of the signal, and the second part $a_m(t)$ represents the effect of the multiples (Fig. 4.51). The desired deconvolution operator $g(t)$ is then such that:

$$a(t) * g(t) = a_0(t) \tag{4.40}$$

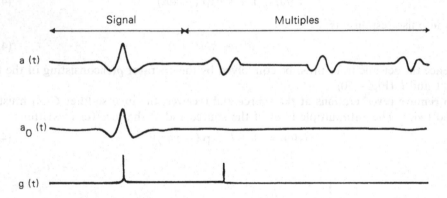

Fig. 4.51 Predictive deconvolution;
  $a(t)$ = autocorrelation of seismic trace with estimation of signal autocorrelation and multiples autocorrelation,
  $a_0(t)$ = signal autocorrelation,
  $g(t)$ = deconvolution operator.

One operator is generally determined per trace, in order to take account of the variations of the multiples from one trace to another, due to the variations in the water depth, for example.

Figure 4.52 shows an example of predictive deconvolution. It shows a marine seismic section in the North Sea, where the signal is deformed by the reverberations from the water layer and contains at least three successive oscillations. After deconvolution, the signal is adjusted to a short pulse allowing the clear distinction of the pinchouts located below the unconformity around 2.2 s.

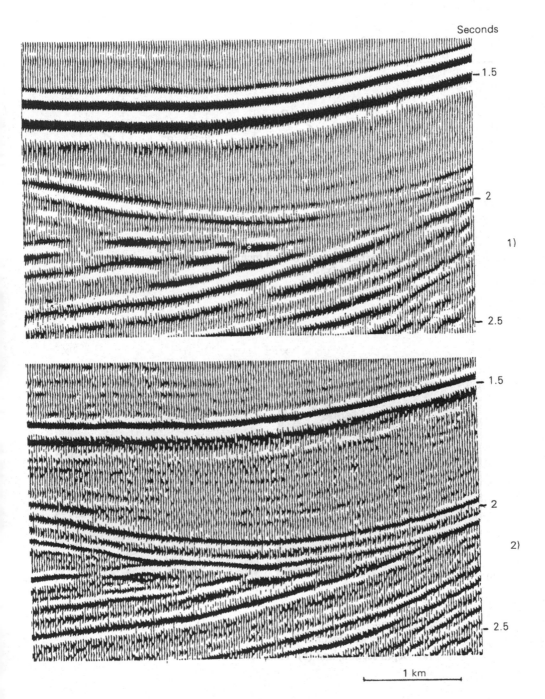

Fig. 4.52 Example of deconvolution after stacking (predictive method). Marine seismic section (*IFP* Document).

(1) Before deconvolution, and (2) After deconvolution.

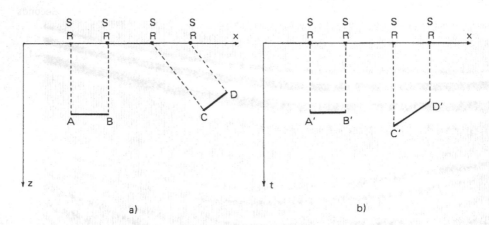

Fig. 4.53 Horizontal reflector *AB* and inclined reflector *CD* in depth section (*x*, *z*) and time section (*x*, *t*).

(a) Depth section, and **(b)** Time section.

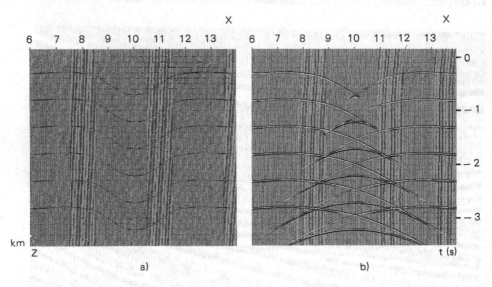

Fig. 4.54 Synclines in depth section (*x*, *z*) and time section (*x*, *t*). (*Prakla-Seismos* Document).

(a) Depth section, and **(b)** Time section.

4.2.10 Migration

As we showed in Section 4.2.8, the CMP stacked section is an approximation of the section that would be obtained by placing the source and detector at the CMP, with the seismic raypaths returning on themselves after normal reflection from the reflectors. It is easy to observe that, if the reflectors are dipping, they are not represented in their true spatial position in the CMP stacked section. In Fig. 4.53, consider the horizontal reflector AB and the dipping reflector CD, shown as a function of depth z, and their displays $A'B'$ and $C'D'$ as a function of time t. The display as a function of time is the section that would be obtained by placing the sources and detectors at the same CMP. The horizontal reflector AB of section (x, z) appears horizontal in the display (x, t). If the time scales are selected to correspond vertically to the depth scales, $A'B'$ even merges with AB. The dipping reflector CD of section (x, z), however, appears at $C'D'$ in the display (x, t), strongly shifted downdip, and its display is therefore incorrect. Anticlines and synclines also give incorrect representations in the display (x, t). Figure 4.54 shows that synclines may assume the appearance of anticlines, if their radius of curvature is smaller than that of the incident wavefront. Also observable are singular reflection points due to the presence of several possible seismic paths.

The purpose of migration is essentially to replace the dipping reflections in their true spatial position on section (x, t). After migration is done, the passage to the depth section is then simply a matter of vertical scale dilatation, an operation that is easy if the average velocities are known. Several techniques are available to replace the dipping reflections in true position, and we shall discuss only two of them, the method of summing up along diffraction curves and the wave equation method.

4.2.10.1 Method of summing amplitudes along diffraction curves

In this method, the reflectors are considered as an infinity of juxtaposed diffracting points. Reflector CD in Fig. 4.55, for example, is considered an infinity of M diffracting points. It is assumed for convenience that a vertical scale dilatation has been carried out, in

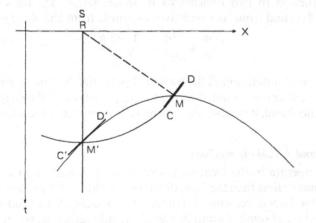

Fig. 4.55 Migration by summing up along diffraction curves.

order to adjust the time and depth to identical scales. It can be shown (Hagedoorn, 1954) that the seismic arrivals M' corresponding to diffraction point M are located on a diffraction curve (a hyperbola if the velocity is constant) which takes the diffracting point M as the apex and the reflector $C'D'$ as tangent at M'. The desired diffracting point M is hence the apex of the diffraction curve tangent at M' to the reflector $C'D'$. Migration by summing amplitudes along diffraction curves thus consists in gathering the seismic arrivals from section (x, t) along each of the diffraction curves defined by the velocity law, summing them, and positioning them at the apex of each hyperbola. The operation is performed automatically by the computer, provided velocity law and stacking windows are defined. This method presents advantages and drawbacks: it is rapid, and it can take account of possible lateral velocity variations, and even carry out migrations in three-dimensional space. On the other hand, it is liable to deform the seismic signals and raise the noise level.

4.2.10.2 Wave equation migration

This method is based on the principle of the downward continuation of the wave field observed at the surface (Claerbout, 1976). Roughly speaking, it consists in progressively lowering the source/recording surface by computation, in order to approach the reflector. It can be shown that whenever the source/recording surface reaches the level of a reflectors, this reflector is migrated. In fact, Fig. 4.53 clearly shows that the closer the source/receiver pairs to the reflectors, the less the reflectors are shifted in (x, t) display. Finally, when the source/receiver pairs reach the level of the reflectors, the reflectors are in the migrated position on section (x, t).

To summarize the operation, the recordings that would be obtained with source/receiver pairs at increasing depths are computed, and the early part of these recordings is preserved. This computation is related to the downward continuation of a wavefield. Several techniques are employed.

a. Finite difference method

The potential $\Phi(x, z, t)$ is discretized with sample intervals Δx, Δz, Δt, assuming the problem to be limited to two dimensions in space. $\Phi(z + \Delta z)$, for example, can be computed from $\Phi(z)$ and from its derivatives obtained from the wave equation:

$$\frac{\partial^2 \Phi}{\partial x^2} + \frac{\partial^2 \Phi}{\partial z^2} = \frac{1}{V^2} \frac{\partial^2 \Phi}{\partial t^2} \tag{4.41}$$

This leads to the computation of the wavefield potential Φ at increasing depths z. The method offers the advantage of not affecting the amplitudes and the shape of the seismic pulses. On the other hand, it is relatively expensive in computer time, especially with high dips.

b. FK analyses and Kirchhoff methods

FK analyses operate in the frequency/wavenumber domain and are generally less expensive in computer time than the finite difference methods, but present the drawback of not accounting for lateral velocity variations. The Kirchhoff method is an analytical variant of the method of summing amplitudes along diffraction curves. It is ideal for high dips, but also presents difficulties for the introduction of lateral velocity variations.

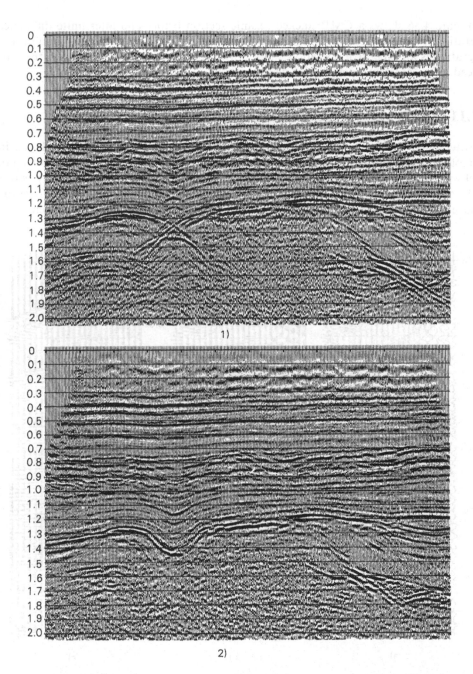

Fig. 4.56 Migration by wave equation, finite difference method. CMP stacked section (*Prakla-Seismos* Document).

(1) Before migration, and (2) After migration.

Figure 4.56 shows a CMP stacked seismic section before and after finite difference migration. Note how the syncline located at depth at 1.45 s is restored by migration. Migration is an expensive operation, but it is virtually indispensable as soon as the dips exceed a few degrees, especially for deep horizons.

4.2.11 Final display of seismic sections

After CMP stacking, or after migration, seismic traces are generally displayed in variable area (see Figs 4.56 and 4.57). The traces are juxtaposed in order to form a raster of tracks whose origins are placed as the source/receiver midpoints in the field. The positive peaks are darkened, while the troughs are left white. The two-way traveltimes from the surface to the reflectors are displayed vertically, with a time scale of 5 or 10 cm/s, or

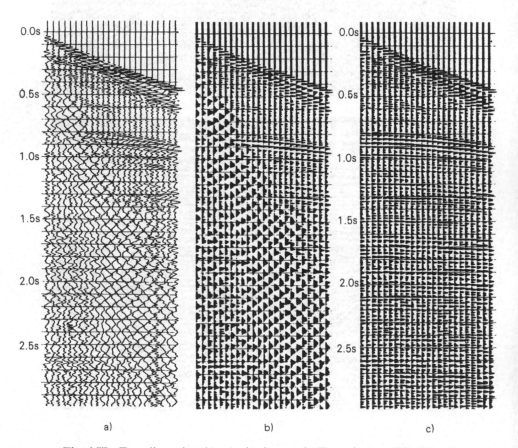

a)   b)   c)

Fig. 4.57 Two-dimensional (x, t) seismic records. (From Anstey, 1981, *Seismic Prospecting and Instrument Specification*, Vol. 1, Gebrüder Borntraeger, Berlin).

(a) Wiggle traces; **(b)** Variable area traces, and **(c)** Variable area traces after filtering for attenuation of surface waves.

20 cm/s for a high resolution survey. The horizontal scale often corresponds to 25 or 50 m between traces, 5 or 10 m in a high resolution survey.

Another display often employed is the **combined mode display** in which the positive peaks are darkened as in the variable area display, and the troughs drawn by wiggle traces (see Figs 4.46 and 4.47). This retains all the signal features. This display is often employed for detail surveys, especially when seeking information about the stratigraphic and lithological variations, or when attempting to fit a seismic survey to the wells.

4.2.12 Three-dimensional processing

Three-dimensional processing involves all the conventional processing operations, plus supplementary processing that we shall analyze briefly.

a. Marine surveys

It is necessary to check that the desired CMP coverage is always satisfied, especially in areas where imprecision of navigation and streamer drift may make the coverage uncertain. Processing therefore begins by the computation of the CMP coverage of each common midpoint, from the effective position of the shots and the detectors. If certain portions of the CMP grid are insufficiently covered, data acquisition must be repeated to fill the gaps. The preliminary processing ("binning") must therefore be carried out on line, during the survey, in order to resume the poorly-covered areas without delay.

b. Land surveys

Static corrections must be carried out for a much larger number of source/detector pairs in 3D, and use can be made of the data redundancy due to the large number of sources and detectors. For **velocity analyses** and **NMO corrections,** special programs must be run which take account of the source geophone azimuth. **3D migration** should theoretically be carried out by gathering the seismic arrivals along the diffraction hyperboloids, summing them and positioning them at the apex of each of the hyperboloids, or by using three-dimensional wave equation methods (finite differences, FK or Kirchhoff).

These very expensive methods are often replaced by the so-called **2 × 2D migration** method, which consists in carrying out a first 2D migration in one direction, such as that of the detector lines, followed by a second migration in the perpendicular direction. It can be shown that 2 × 2D migration generally yields results close to the true 3D migration. The character of the seismic reflections may be altered in some cases.

The **final display** of a 3D survey is in the form of a set of sections across a 3D block, namely vertical sections in selected directions and horizontal sections at selected depths (Fig. 4.58). This allows a very complete interpretation of the subsurface structures.

Interpretation is generally carried out on graphic **interactive workstations,** designed for the simultaneous or consecutive display of several vertical and horizontal sections (Fig. 4.59). Interactive mode allows the modification of certain parameters, such as the velocities or the horizon picking, to obtain a coherent image of the subsurface and to plot isobath maps. Graphic interactive workstations, combined with expert systems, are probably destined for considerable growth in the years to come, for the processing and interpretation of three-dimensional seismic surveys.

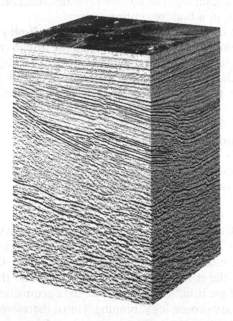

Fig. 4.58 Three-dimensional seismic display: vertical sections in two perpendicular directions and horizontal section. (*Geco* Document).

Fig. 4.59 "Interpret" interactive interpretation work station. Selection of functions on a menu with electric pen. (*CGG* Document).

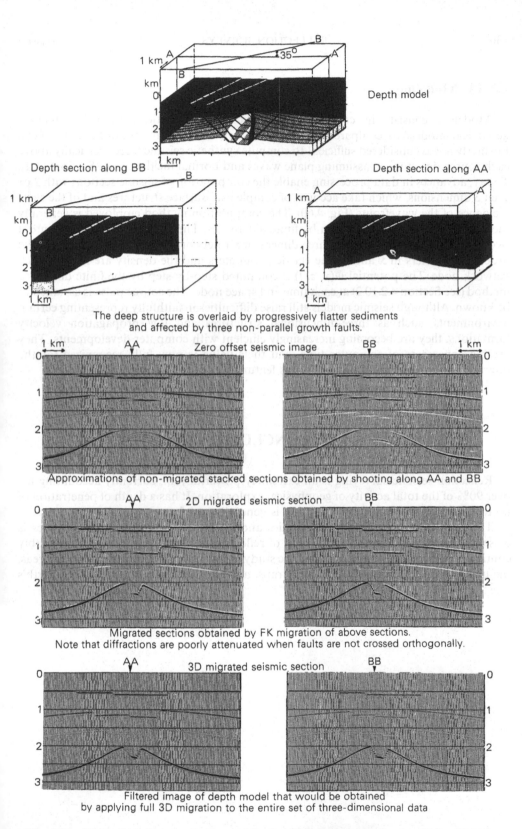

1 km   A   B   135°   A

km 0 1 2 3   1 km   Depth model

Depth section along BB   Depth section along AA

The deep structure is overlaid by progressively flatter sediments
and affected by three non-parallel growth faults.

1 km   AA   Zero offset seismic image   BB   1 km

Approximations of non-migrated stacked sections obtained by shooting along AA and BB

AA   2D migrated seismic section   BB

Migrated sections obtained by FK migration of above sections.
Note that diffractions are poorly attenuated when faults are not crossed orthogonally.

AA   3D migrated seismic section   BB

Filtered image of depth model that would be obtained
by applying full 3D migration to the entire set of three-dimensional data

Fig. 4.60   Three-dimensional seismic model of a faulted structure in the Gulf of Mexico.
(*CGG* Document).

4.2.13 Modeling

Modeling consists in computing synthetic seismic sections from hypothetical geological models to compare them with the seismic sections obtained in the field. Formerly it was considered sufficient to compute synthetic seismic traces vertically above each source/detector set, assuming plane waves and normal incidence (see Section 3.6.1). Today, advances in data processing enable the computation of seismic sections with 2 or even 3 dimensions, which take account of complex subsurface structures and of the slant incidence of the wavefronts (Fig. 4.60). The computation method employed is generally the solution of the wave equation by finite differences. The subsurface is divided into a two-dimensional (and possibly three-dimensional) network, with an increment on the order of 10 meters in x and z. The elastic parameters and the density are fixed at each network node. The potential $\Phi(x, z, t)$ is computed step-by-step by the finite difference method (see Section 4.2.10.2) at each time and space node. The source pulse is assumed to be known. Although seismic models still raise difficulties in faithfully representing certain environments, such as three-dimensional structures and low propagation velocity formations, they are becoming increasingly efficient with computer developments. They are big consumers of computer time, and their value is now fully recognized for the interpretation of seismic surveys in areas featuring complex tectonics.

4.3 CONCLUSIONS

Reflection seismic surveying is an extremely powerful tool, currently accounting for over 90% of the total activity of geophysical exploration. It has a depth of penetration of up to 10 km or more, and its resolution is generally satisfactory.

The tremendous progress in electronics and computer science in recent decades is closely linked to the spectacular growth of reflection surveys. This trend will probably continue in the years to come, allowing the study of increasingly complex structural areas, the determination of new stratigraphic traps, and the deep investigation of the earth's crust.

5

transmission surveys
and vertical seismic profiles

Transmission surveying examines the direct propagation of seismic waves between sources and detectors placed in boreholes and galleries, to obtain data on the acoustic properties of the formations. It is mainly employed in mining and civil engineering. Vertical seismic profiles (VSPs) are designed to analyze the propagation of seismic waves between the surface and detectors positioned in wells. VSP is often conducted by oilmen to obtain information about the acoustic properties of the formations around the wells.

5.1 CHANNEL WAVE TRANSMISSION

Channel waves are sometimes used in coalmines to test the continuity of the seams. The propagation velocities in coal are around 2500 m/s for P-waves, 1200 m/s for S-waves, and the density is 1.6. These values are significantly higher in the surrounding sediments. Coal beds therefore often act as waveguides, with the seismic energy propagating along the seams in the form of dispersive waves, similar to pseudo-Rayleigh and Love waves along the free surface.

Dispersion analysis provides information about possible interruptions in the seams ahead of the coal front. Figure 5.1 shows a transmission survey between two galleries with one shotpoint and 24 detection points, and the recording obtained (Millahn, 1980). The first arrival observed is that of P-waves around 40 ms, followed by S-waves around 75 ms, and finally the channel waves. The P- and S-waves propagate in the surrounding formations at 4200 and 2300 m/s respectively. The Love waves exhibit a dispersive character, with separation of the phase and group velocities (Fig. 5.2). The low frequencies (150 to 200 Hz) arrive first, at velocities of about 2300 m/s, next the higher frequencies (300 to 400 Hz) at lower velocities, and finally the Airy phase, at high frequency (470 Hz) and

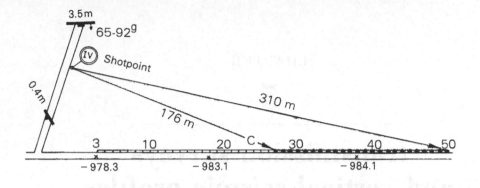

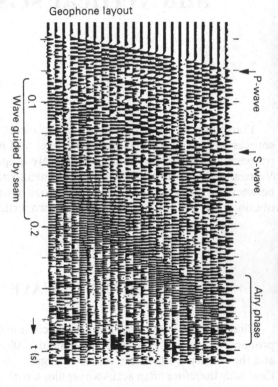

Geophone layout

Fig. 5.1 Plan view of seismic transmission survey between mine galleries and recording obtained.
(*Prakla-Seismos* Document)

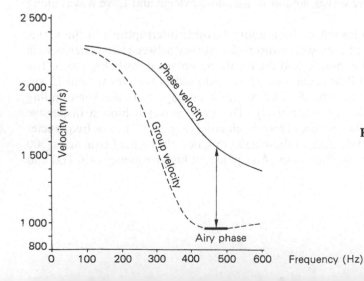

Fig. 5.2 Dispersion curve of channel waves, for Fig. 5.1 (Love-type waves, fundamental mode).
(*Prakla-Seismos* Document).

low velocity, group velocity 1000 m/s, phase velocity 1600 m/s. The Airy phase frequency depends on the seam thickness h by the formula:

$$\frac{2\pi f h}{c} = k \tag{5.1}$$

where

c = phase velocity,

k = a constant which depends on the walls, about 3 in this case.

The thinner the seam the higher the frequency. The frequency of 470 Hz and the phase velocity of 1600 m/s indicate a thickness of about 1.6 m. The fact of receiving an energetic Airy phase indicates the absence of any interruption in the seam between the source and the detectors. If the seam were intersected by a fault, the position of the fault could be analyzed by reflection, by positioning detectors in the same gallery as the sources.

Channel waves can propagate over distances of about 1000 times the seam thickness in favorable cases. They are extremely valuable for predicting mining conditions.

5.2 VELOCITY TOMOGRAPHY BY TRANSMISSION

Tomographic analyses can be conducted by transmission from borehole to borehole or from gallery to gallery (Bois *et al.*, 1972, La Porte *et al.*, 1973, Ivansson, 1985). The seismic velocities are determined between the boreholes or the galleries by computing the raypaths and by iterative interpretation of the propagation times (Fig. 5.3). Iterative interpretation consists in selecting an initial velocity model, computing the propagation times across this scheme, comparing them with the experimental times, adjusting the initial scheme, and so on, until the difference between the computed and measured velocity tomograms is sufficiently small. Velocity tomograms can be computed for compressional waves and for shear waves. Amplitude and attenuation tomograms, and acoustic impedance tomograms can also be computed.

Transmission tomograms help to improve the understanding of the subsurface by detecting major heterogeneities between the boreholes and the galleries. They provide valuable information about the mechanical properties of the formations in mining and in civil engineering. Transmission surveying is employed today in mining and civil engineering, but less in petroleum exploration. In the future, however, it can be expected to provide useful data for the detailed investigation of reservoirs.

5.3 VERTICAL SEISMIC PROFILING

VSP consists in generating a signal at the ground surface and in detecting the seismic arrivals in a well, with geophones anchored at different depths (Galperin, 1973, Wuenschel, 1976, Hardage, 1983). **First arrival well shooting** recordings were generally

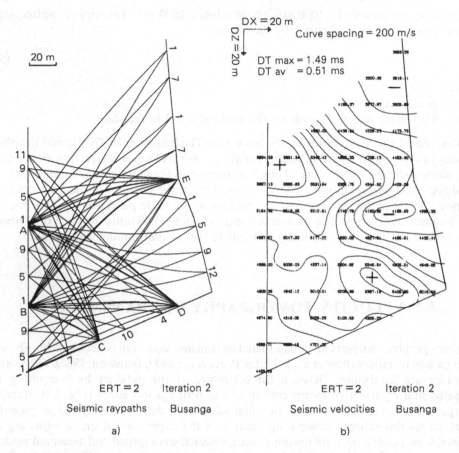

Fig. 5.3 Plan view of seismic transmission survey between exploration galleries for locating an underground hydropower plant. (From La Porte *et al.*, 1973, "Mesures sismiques par transmission, application au Génie Civil", *Geophysical Prospecting*, Vol. 21, No. 1).

(a) Automatic raypath plot, and (b) Tomogram of velocities obtained, A, B, C, D and E: shotpoints. Figures: geophones.

considered adequate before 1976. These operations measured the average velocity as a function of depth, by lowering a geophone at widely-spaced depths into a hole, at 300 to 400 m intervals, for example, and recording energy from shots fired from the surface. In VSP, however, recordings are made with much closer detection points, at 10 or 20 m intervals, for example, and, in addition to the first arrivals, the various ascending and descending waves following the first arrivals are also interpreted. The well may be several thousand meters deep, and a well geophone (or an array of several geophones) is lowered and anchored against the well wall, at intervals ranging from a few meters to 20 or so. The VSP is obtained by emitting a seismic pulse at the surface, and by recording the succession of seismic traces for all the depths of the geophone in the well (Fig. 5.4).

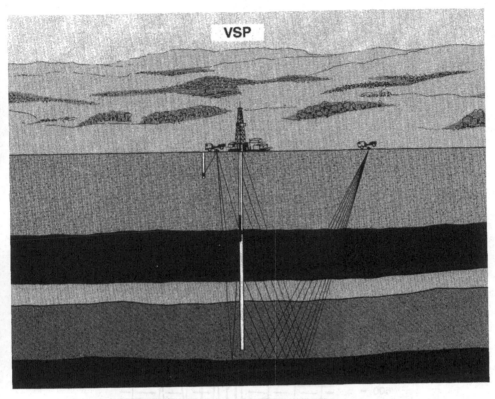

Fig. 5.4 Vertical seismic profile and offset vertical seismic profile. (*Prakla-Seismos* Document).

Left : VSP.
Right : Offset VSP.

If the source is located near the well (Fig. 5.5), VSP helps to estimate the reflectivity of the different reflectors, the deformation of the pulse during propagation, and the relative importance of secondary reflections. Small-offset VSP is an excellent tool for optimizing processing in reflection surveying, and can help to improve the removal of multiples or to optimize signal deconvolution. If the source is located some distance from the well, a few hundred meters or more (Fig. 5.6), VSP is called offset VSP. It helps to obtain information about the structures located in the neighborhood of the well, such as domes, faults and reefs, to determine their boundaries in space, and to analyze their degree of fracturing.

5.3.1 Emission

The sources are generally the same as in reflection surveying. Vertical vibrators and weight drops are used for *P*-waves. Air guns and water guns can also be placed in the mud pits near the wellhead. Horizontal vibrators and impactors are employed for *S*-waves.

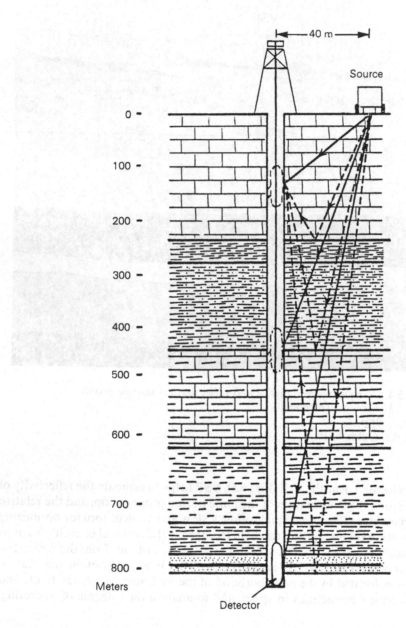

Fig. 5.5 Small offset VSP.

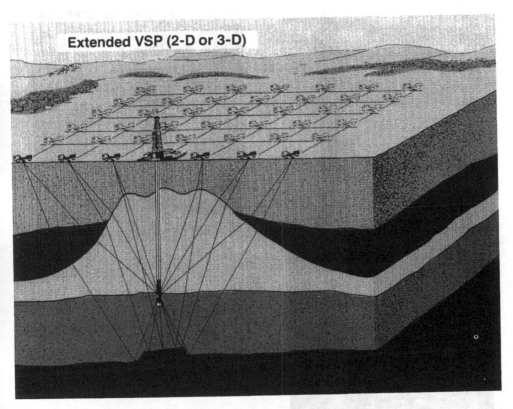

Fig. 5.6   Three-dimensional extended VSP to delineate a structure around the well. (*Prakla-Seismos* Document).

Offshore sources are usually air guns or water guns. The emission time and the seismic signature are recorded by a reference geophone or hydrophone, positioned at a depth of some 30 m below the source.

5.3.2 Detection

A detector generally consists of a geophone or a small geophone group, positioned vertically in a well probe, or a three-directional array positioned as a trirectangular trihedron in the probe. The probe often contains a multiplexer and an analog-to-digital converter, to transmit the data to the surface in digital form, with a better signal-to-noise ratio. It is generally coupled to the well wall by an anchoring system (Fig. 5.7 a) enabling the geophones to pick up the particle motion in the ground accurately, in a sufficiently wide passband, 0 to 250 Hz for example. An array of several superimposed probes can be used to facilitate implementation.

(b)

Fig. 5.7 *CGG/IFP* (*ARTEP*) SPH probe. (*CGG* Document).

(a) Probe and anchoring system.
(b) Recording system in recording truck.

(a)

5.3.3 Recording

The recording system is approximately the same as in reflection surveying (see Section 4.1.4), with the possibility of carrying out preprocessing in the field, such as the stacking of a large number of emissions (Fig. 5.7 b). Figure 5.8 shows a raw VSP recording of *P*-waves before processing. A first arrival can be seen, the descending direct wave, and reflected arrivals ascending after reflection from the subsurface discontinuities. Also observable are descending secondary arrivals, parallel to the direct arrival, which are secondary reflections, reflected first upward and returned downward. The mechanism of the formation of ascending and descending reflections is shown in Fig. 5.9.

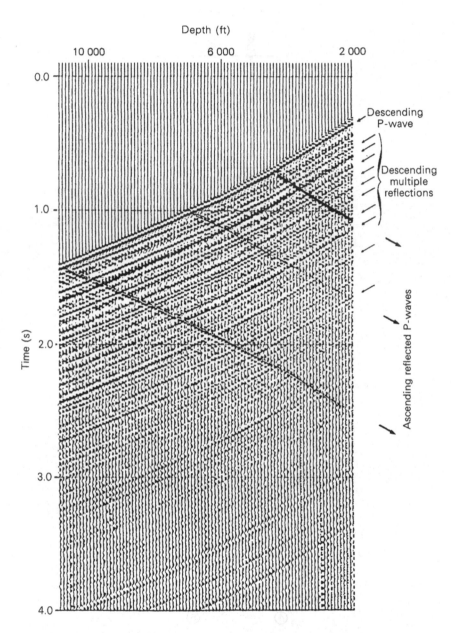

Fig. 5.8   Raw VSP record. (*SSL* Document).

Descending wave

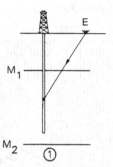

Ascending reflections

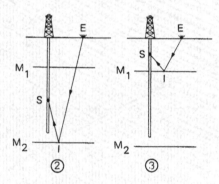

Descending multiple reflections

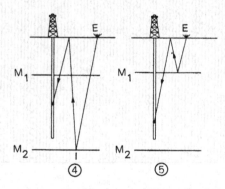

Fig. 5.9 Paths of descending waves, ascending reflections, descending multiple reflections. (*CGG* Document).

Many raw VSP recording are cluttered by spurious waves, particularly by fluid waves, which propagate vertically in the mud of the well at velocities around 1500 m/s (Fig. 5.10). These are coupled mud formation waves, slightly affected by the formation, which offer little information about the formation velocities. They may, however, be sensitive to certain petrophysical properties of the formations.

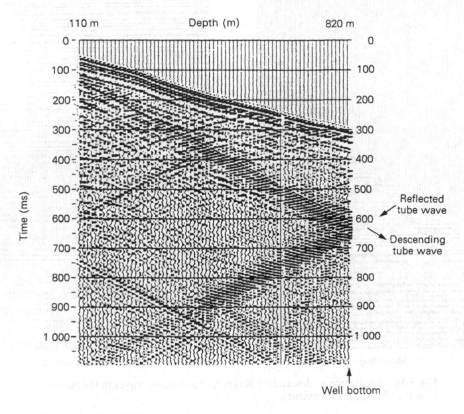

Fig. 5.10 Raw VSP record of *P*-waves disturbed by the tube wave. (*CGG* Document).

5.3.4 Processing

VSPs, formerly processed at the central lab, are now increasingly processed in the field with the application of suitable filters to remove the fluid waves and to separate the ascending and descending waves (Fig. 5.11). The ascending waves serve to identify single reflections, and the descending waves help to analyze the signal deformations in the subsurface and to identify secondary reflections. By applying a downward shift of t_0 to the ascending waves, and an upward shift of t_0 to the descending waves, where t_0 denotes the time of the first arrival, the ascending arrivals and descending arrivals are distinguished, and the VSP can be fitted to the surface reflection recordings (Fig. 5.12).

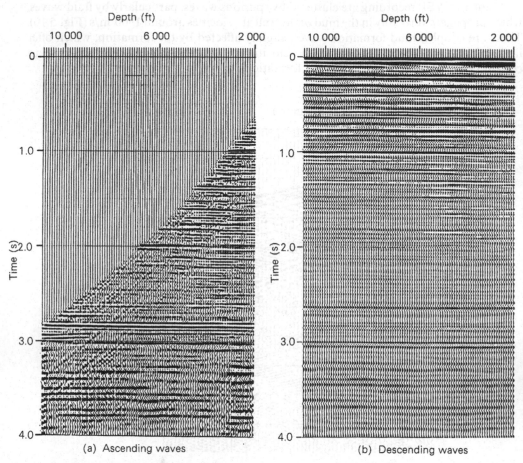

Fig. 5.11 Separation of descending waves and ascending waves in the record in Fig. 5.8. (*SSL* Document).

Inversion of the VSP serves to compute the acoustic impedances and to obtain a forecast of the impedance logs in areas not yet reached by the drill (Fig. 5.13).

VSP can also be carried out in *S*-waves, by using *SH*-wave sources and horizontal geophones (Fig. 5.14). Simultaneous *P*- and *S*-waves VSPs help to compute the ratio of dilatational to shear velocities V_p/V_s, and to determine certain lithological and petrophysical properties of the formations. If *P*-wave and *S*-wave seismic profiles are available around the well, the *P*- and *S*-sections can be fitted to each other accurately through VSP. This yields new information about the subsurface geology in the areas covered by the profiles (see Section 4.1.8).

VSPs are extremely valuable for the calibration of surface seismic data on wells, the determination of processing parameters, and the identification and removal of secondary reflections. They are increasingly used for the investigation of the structures around the wells, for the detection of fractures, and for assessing the petrophysical properties of the reservoirs.

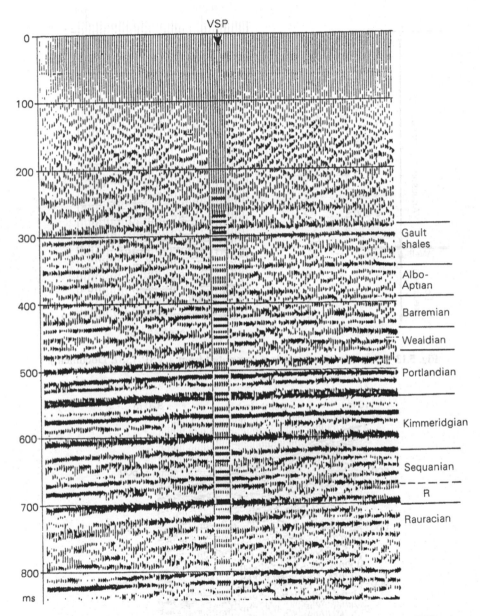

Fig. 5.12 Calibration of a reflection seismic section to a VSP. (*IFP* Document).

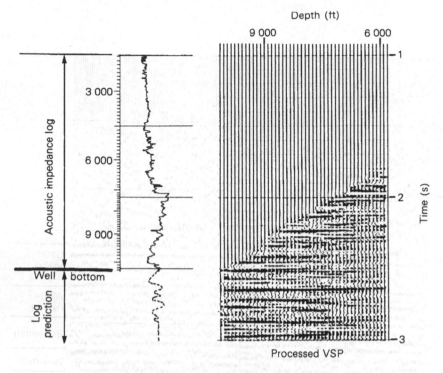

Fig. 5.13   Acoustic impedance log, computed from a VSP. (*SSL* Document).

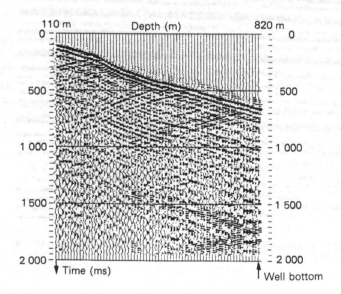

Fig. 5.14   *S*-wave VSP obtained in the well in Fig. 5.10. (*CGG* Document).

6

refraction surveys

Refraction surveying is an exploration method that consists in observing at the surface the head waves, refracted along high velocity interfaces, in order to determine the geological structures. Interfaces capable of constituting "markers" in refraction shooting are those separating an upper medium of velocity V_1 and a lower medium of velocity V_2 higher than V_1. Roughly speaking, it can be considered that the head waves travel obliquely at velocity V_1, downward from the source, enter the high-velocity medium at the critical angle:

$$\theta_L = \text{arc sin } \frac{V_1}{V_2} \tag{6.1}$$

travel along the interface at velocity V_2, emerge at angle θ_L in the upper medium, and travel obliquely upward to the ground surface at velocity V_1 (Fig. 6.1).

The method attempts to map the different refractors in depth and to determine the propagation velocities. This is done by the accurate determination of the head wave arrival times at the ground surface as a function of the source/detector distance. It is generally necessary to fire "direct" shots from S to geophones R_1, R_2, ... R_n and "reverse" shots from S' to geophones R_n, R_{n-1}, ... R_1, to be able to compute the depths, velocities and dips simultaneously.

Beyond a certain source/detector distance, around twice the depth of the refractors or so, the head waves reach the detectors first. When the source/detector distance increases, the head waves come from increasingly deep and increasingly high-velocity refractors (Fig. 6.2). Note that a high-velocity layer can only serve as a refractor if it is continuous and sufficiently thick; otherwise, the head wave is attenuated during horizontal propagation and cannot be detected at long distances.

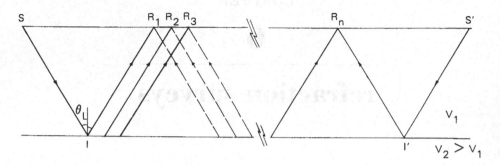

Fig. 6.1 Onshore refraction survey. From S to $R_1, R_2, \dots R_n$ direct shot. From S' to $R_n, \dots R_1$ reverse shot.

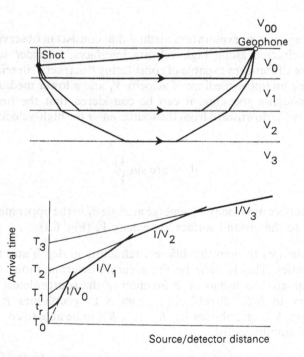

Fig. 6.2 Raypaths and arrival times of head waves, as a function of source/detector distance, for three horizontal layers overlying the basement. (From Musgrave, 1967, *Seismic Refraction Prospecting*, Society of Exploration Geophysicists, Tulsa, Oklahoma, USA).

6.1 DATA ACQUISITION

a. Source-detector spreads

The main difference from reflection surveys is the long source/detector distance. This can reach 60 km in refraction surveys, compared with a few kilometers in reflection surveys. This means that the energy emitted by the source must be much greater, and the detonation of a few hundred kilograms of explosive in refraction shooting is not a rare occurrence.

b. Frequency band

The long refraction paths cause attenuation of the high frequencies, and refraction recordings are generally at lower frequency than reflection recordings. The recording passband often extends from 4 to 60 Hz, against 10 to 125 Hz in conventional reflection surveying. The geophones are hence at a lower frequency, with a natural frequency of 4 Hz, for example, against 10 or 20 Hz in reflection surveying. Despite the difference in passbands, the digital recording instruments for reflection surveying can generally be used for refraction, with a different setting of the low-cut and high-cut filters.

6.1.1 Land surveys

The principle generally employed is the measurement of direct and reverse propagation times.

a. Long-distance refraction

The **direct shot** is fired at one end of the profile and the head waves are recorded at increasing distances from the source, extending from 5 to 30 km from the source, for example. The reverse shot is fired at the other end. The traces may be spaced at 100 m intervals, with each detector consisting of a pattern of several geophones. Since it is inconvenient to record 200 to 300 traces simultaneously extending over 20 to 30 km, refraction profiles are generally recorded in segments a few kilometers long, each with 20 to 40 traces. The source is often an explosive source. The charge, from a few kilograms for the nearest arrays, are up to 100 kg for the furthest. Large vibrators are also employed.

Long-distance refraction shooting is generally used by oilmen to find the crystalline basement, and by earth physicists to analyze the Earth's crust. Figure 6.3 gives an example of a refraction profile recording, with direct and reverse shots, using five spreads of 1500 m extending from 4500 to 12,000 m (Musgrave, 1967).

b. Weathering shots

Refraction profiles are also employed in petroleum and mining geophysics to determine the thickness and velocities in the weathered zone and the surface layers. If, for example, the weathered zone is 5 to 15 m thick, recordings can be made with small refraction spreads of a few hundred meters. Information is obtained on the thicknesses and velocities in the weathered zone, allowing static corrections in land surveys (see Section

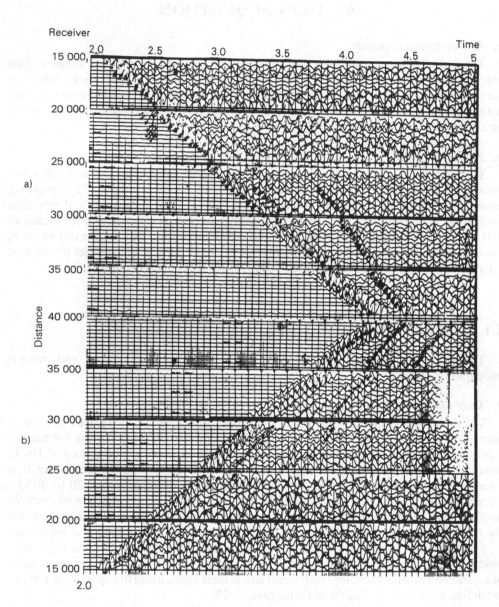

Fig. 6.3 Record of a refraction profile with direct shot and reverse shot. Distances in feet. (From Musgrave, 1967, *Seismic Refraction Prospecting*, Society of Exploration Geophysicists, Tulsa, Oklahoma, USA).

(a) Direct shot, and **(b)** Reverse shot.

4.2.5). Figure 6.4 shows a small refraction recording for the investigation of surface layers (Musgrave, 1967). The shots are fired at the center of a 24-trace seismographic spread extending over 780 m. Several refraction arrivals are distinguished, at velocities between 1320 and 1830 m/s.

6.1.2 Marine surveys

a. Sonobuoy method

A cheap method was formerly used, requiring a single ship, the sonobuoy method. A buoy equipped with a hydrophone, a recorder and a radio transmitter is thrown into the water. The ship travels away in the profile direction while emitting seismic pulses. The seismic recordings received by the hydrophone are transmitted by radio to the ship, to distances up to some 40 km. A refraction survey is rapidly recorded, with direct shots only.

Despite the loss of the buoy (which is often abandoned on completion of the profile), the sonobuoy method is much less expensive than methods requiring two or more ships. However, the measurements are often unreliable, because the currents and winds cause the buoy to drift.

b. Expanding spread profiles (ESP)

Marine refraction surveys are now often conducted with two ships. One fires the seismic shots, and is equipped with a powerful source, such as an array of several air guns. The second performs detection, towing a 1200 to 2400 m streamer, with digital recording instrumentation onboard.

The two ships, some 50 km apart at the start of the operation, travel towards each other at a velocity of 5 knots, shooting and recording every 20 s, for example. They cross and then travel away from each other some 50 km, always following the predetermined profile, while the midpoint between the ships remains invariable. The reciprocal time principle implies data redundancy before and after crossover. The accuracy of navigation and positioning can be checked by comparing the symmetrical recordings with respect to the crossover point, which should be identical for two symmetrical positions of the ships. Figure 6.5 shows an expanding spread profile obtained with two ships. The extreme right-hand trace corresponds to a distance of 50 km between the ships, and the extreme left-hand trace to 1 km. One trace is recorded whenever the distance between the ships decreases by 100 m. The symmetrical traces are not shown. Recording begins 7 s after the shot and lasts 10 s. A reflection curve on the seafloor is easily distinguished. The vertical two-way time of 7.35 s indicates a water depth of 5500 m. The very energetic seafloor/surface multiple reflection curve is also observable.

The first observed arrivals from 6000 to 32,000 m, at times ranging between 8.8 and 12.7 s, are a set of waves at velocities of 6500 to 7000 m/s, refracted from a high-velocity refractor, possibly the crystalline basement below the ocean floor. A second relatively energetic arrival extends from 23,000 to 35,000 m, at times between 12 and 13.3 s, with a velocity of about 8000 m/s. This probably represents a head wave refracted from a deeper and even higher velocity refractor, possibly the basaltic formation. A third rectilinear arrival can be seen from 10,000 to 32,000 m, at times between 10.9 and 17 s, at a velocity of about 3600 m/s. This could correspond to a head wave refracted along a sedimentary layer.

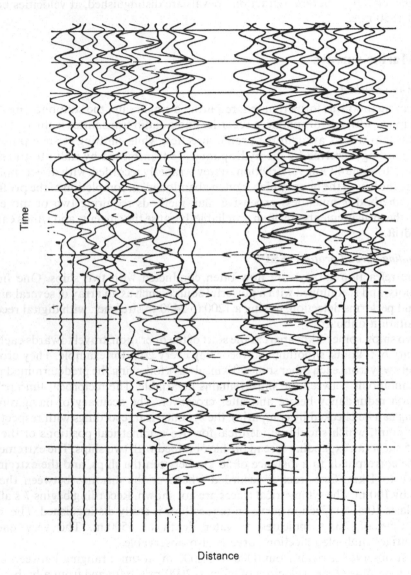

Distance

Fig. 6.4 Refraction record for the investigation of surface layers. Shot in the center. 24 traces, 30 m between traces, 50 ms between time lines. (From Musgrave, 1967, *Seismic Refraction Prospecting*, Society of Exploration Geophysicists, Tulsa, Oklahoma, USA).

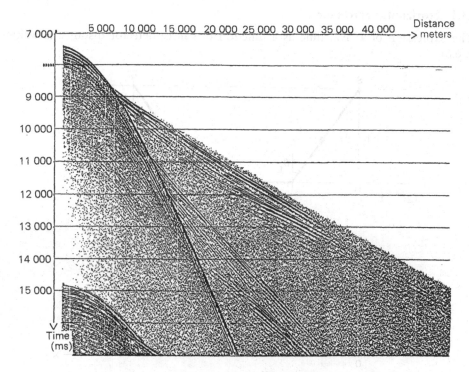

Fig. 6.5 Record of expanding spread profile (ESP) with common midpoint obtained with two ships in a water depth of 5500 m. (*IFP* Document).

The expanding spread profile does not involve the measurement of reciprocal propagation times, contrary to appearances. It is therefore difficult to determine the velocities and depths accurately in the presence of variable dips.

6.1.3 Signals received in refraction shooting

For the sake of simplicity, we shall restrict ourselves to monocline structures taking the vertical plane of the profile as the plane of symmetry. Let us assume that a refractor exists, horizontal or inclined, between two media of velocities V_1 and $V_2 > V_1$, and let:

$$\theta_L = \text{arc sin } \frac{V_1}{V_2}$$

be the critical angle. Head waves refracted at the critical angle θ_L travel along the refractor at velocity V_2, merge in the upper medium, and arrive at the free surface along which they travel at an apparent velocity V_2 (Pilant, 1979).

We shall examine the time/distance curves obtained with a horizontal refractor and an inclined refractor.

6.1.3.1 Horizontal refractor

The expression of arrival time as a function of source/detector distance is written (Fig. 6.6):

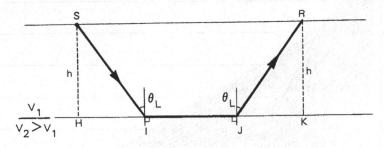

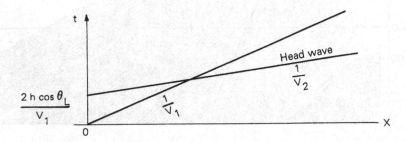

Fig. 6.6 Raypaths and time distance curves for a horizontal refractor. Direct shot.

$$t = \frac{SI + JR}{V_1} + \frac{IJ}{V_2} \tag{6.2}$$

$$t = \frac{2h}{V_1 \cos \theta_L} + \frac{x - 2h \tan \theta_L}{V_2} \tag{6.3}$$

where

t = arrival time,
x = source/detector distance,
h = refractor depth,
θ_L = critical angle.

Noting that:

$$V_2 = \frac{V_1}{\sin \theta_L}$$

(6.3) is written:

$$t = \frac{x}{V_2} + \frac{2h}{V_1} \cos \theta_L \tag{6.4}$$

(6.4) is the equation of the time/distance curve of the horizontal refractor.

The ordinate at the origin:

$$t_0 = \frac{2h}{V_1} \cos \theta_L$$

is called the **intercept**. Knowing V_1 and V_2, it serves to compute the refractor depth h. Velocities V_1 and V_2 can be measured directly on the time/distance curves of the surface arrival (V_1) and the head wave arrival (V_2). The half-intercept $h/V_1 \cos \theta_L$ is called the "delay time". It represents the additional time taken by the wave to travel path SI instead of path HI (Fig. 6.6).

6.1.3.2 Dipping refractor

For a dipping refractor with dip α, the expression of arrival time as a function of source/detector distance, for a direct shot (S-wave emission) (Fig. 6.7) is written:

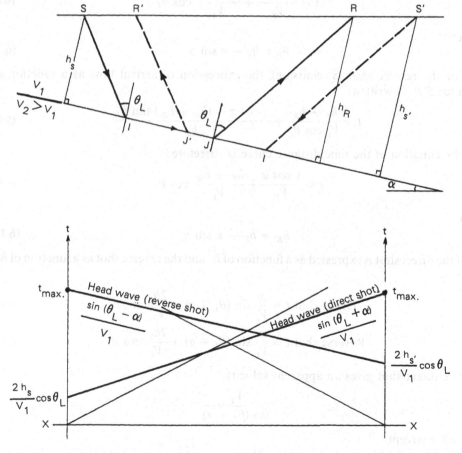

Fig. 6.7 Raypaths and time distance curves for a dipping refractor with dip α. Direct shot and reverse shot.

$$t = \frac{SI + JR}{V_1} + \frac{IJ}{V_2}$$

$$t = \frac{h_S + h_R}{V_1 \cos \theta_L} + \frac{x \cos \alpha - (h_S + h_R) \tan \theta_L}{V_2} \tag{6.5}$$

where

t = arrival time,

x = source/detector distance,

h_S = refractor depth measured perpendicular to the refractor below the source point,

h_R = refractor depth measured perpendicular to the refractor below the detection point,

θ_L = critical angle,

α = refractor dip.

The equation of the time/distance curve is:

$$t = \frac{x \cos \alpha}{V_2} + \frac{h_S + h_R}{V_1} \cos \theta_L \tag{6.6}$$

with

$$h_R = h_S + x \sin \alpha \tag{6.7}$$

For the reverse shot (S' emission), the expression of arrival time as a function of distance $S'R'$ is written:

$$t = \frac{h_{S'} + h_{R'}}{V_1 \cos \theta_L} + \frac{x \cos \alpha - (h_{S'} + h_{R'}) \tan \theta_L}{V_2} \tag{6.8}$$

The equation of the time/distance curve is therefore:

$$t = \frac{x \cos \alpha}{V_2} + \frac{h_{S'} + h_{R'}}{V_1} \cos \theta_L \tag{6.9}$$

with

$$h_{R'} = h_{S'} - x \sin \alpha \tag{6.10}$$

If the direct shot is expressed as a function of h_S and the reverse shot as a function of $h_{S'}$, we have:

$$\text{Direct shot } t = \frac{x}{V_1} \sin (\theta_L + \alpha) + \frac{2h_S}{V_1} \cos \theta_L \tag{6.11}$$

$$\text{Reverse shot } t = \frac{x}{V_1} \sin (\theta_L - \alpha) + \frac{2h_{S'}}{V_1} \cos \theta_L \tag{6.12}$$

The direct shot gives an apparent velocity:

$$\frac{V_1}{\sin (\theta_L + \alpha)}$$

and an intercept:

$$\frac{2h_S}{V_1} \cos \theta_L$$

The reverse shot gives an apparent velocity:

$$\frac{V_1}{\sin (\theta_L - \alpha)}$$

and an intercept:

$$\frac{2h_{S'}}{V_1} \cos \theta_L$$

If V_1 is known, the measurement of the two apparent velocities helps to compute the critical angle θ_L and the dip α. The measurement of the two intercepts serves to determine the depths h_S and $h_{S'}$ measured perpendicular to the refractor below sources S and S'.

It can be seen that the half-sum of the slownesses of the direct and reverse waves:

$$\frac{1}{2}\left[\frac{\sin (\theta_L + \alpha)}{V_1} + \frac{\sin (\theta_L - \alpha)}{V_1} \right] = \frac{\sin \theta_L \cos \alpha}{V_1} = \frac{\cos \alpha}{V_2} \tag{6.13}$$

is not equal to the true slowness $1/V_2$ of the refraction marker, but approaches it with decreasing dip.

The time/distance curves of the refracted wave in direct and reverse shooting are shown in Fig. 6.7 for the dipping refractor. The intercepts and apparent velocities are different, but the maximum time t_{max} to travel from S to S' is the same in direct shooting and reverse shooting.

6.2 PROCESSING AND INTERPRETATION

6.2.1 Preliminary processing and corrections

The preliminary processing of a refraction survey essentially comprises elevation corrections for the source and detection points, and velocity and thickness corrections for the weathered zone.

a. Elevation correction

The source and detection point elevations are adjusted to a suitable datum plane by correcting the corresponding propagation time. Figure 6.8 shows that, for a single refractor, the transit time $SIJR$ must be adjusted to the transit time $S'KLR'$, i.e. by subtracting the times:

$$\frac{SS_1}{V_1} \quad \text{and} \quad \frac{RR_1}{V_1}$$

and adding the times:

$$\frac{KI}{V_2} \quad \text{and} \quad \frac{JL}{V_2}$$

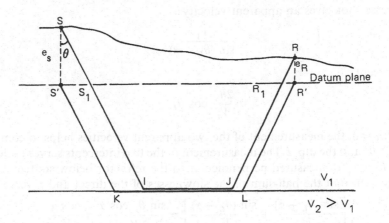

Fig. 6.8 Elevation correction in refraction shooting.

For simplification, let us assume that the refractor and the datum plane are horizontal. If e_s denotes the elevation of the source and e_R the elevation of the detector with respect to the datum plane, the elevation correction is written:

$$C_A = -(e_s + e_R) \frac{\cos \theta}{V_1} \tag{6.14}$$

where

θ = incident angle of the raypaths upon the refractor,

V_1 = velocity in the upper medium.

For two or more interfaces, it can be shown that the elevation correction is given by a similar formula, where θ is the incident angle of the raypaths on the interfaces.

b. Velocity and thickness correction for the weathered zone

As in reflection surveys, the velocity and thickness of the weathered zone must be determined by shots fired at its base, or by small refraction profiles (see Section 4.2.5.1).

6.2.2 Filtering. The tau-p transform

To a certain extent, conventional space/time filterings help to attenuate coherent noises, surface waves or converted waves, and incoherent noises. A more recent and highly effective method is the "**intercept/slowness (tau-p)**" method (Diebold and Stoffa, 1981). The seismic recordings in the time/space domain are converted in the intercept/slowness domain (τ, p) defined by:

$$\text{Intercept } \tau = \frac{2h}{V_1} \cos \theta_L \tag{6.15}$$

$$\text{Slowness } p = \frac{1}{V_2} \tag{6.16}$$

In this transformation, a refracted arrival of velocity V_2, represented in domain (t, x) by a line with equation:

$$t = \frac{x}{V_2} + \frac{2h \cos \theta_L}{V_1} \tag{6.17}$$

is displayed by a coordinate point τ, p (Fig. 6.9).

It can be shown that the reflected arrivals, represented by hyperbolas in the domain (t, x) are displayed by ellipses in domain (τ, p). The organization of noise is generally different in domain (t, x) and in domain (τ, p). Coherent noises, aligned with the lines in domain (t, x) are grouped in clusters of points in domain (τ, p), and it is clear that their removal there is easier. Reverse transformation in domain (t, x) restores a noise-free recording (Fig. 6.10). The improvement of the signal-to-noise ratio then helps to compute the velocities and depths accurately. It is not rare for the tau-p transform to be applied to refraction and wide-angle reflection recordings (near the critical angle), leading to a good assessment of the velocity distribution $V(z)$ as a function of depth.

6.2.3 Interpretation: the Gardner method

Interpretation consists in computing the depth and velocity of the different refractors. It can be carried out in various ways. We shall mention the Gardner method (1939), still used today, which allows manual interpretation or semi-automatic interpretation on a computer.

The intercepts of the **direct shots** are first computed for successive recording segments. Formula (6.6) shows that the intercepts are the sum of the "delay times" at the source and detection points. If the successive intercepts along the profile are plotted against x, the curves obtained display breaks for each source point $S_1, S_2, \dots S_n$, because the source "delay times" are different (Fig. 6.11). These breaks are generally corrected by tying the intercept curves from one segment to the next, by vertical shifting. This serves to reconstruct the intercept curve which would have been obtained if all the shots had been fired at point S_1. The intercepts of the **reverse shots** are then computed and plotted similarly, thus reconstructing the intercept curve that would have been obtained if all the shots had been fired at point S_n.

The fluctuations in the intercept curves obtained in direct shooting and reverse shooting are closely comparable (Fig. 6.12). They reflect the elevation fluctuations of the refractor at points J, and not at points R, and are offset with respect to the refractor by a quantity equal to the horizontal projection of JR, which is called the "offset". To correct this offset, the intercept curve of the direct shot must be shifted horizontally towards S_1, and that of the reverse shot towards S_n. This gives two intercept curves, which reflect the elevation fluctuations of the refractor. The curves are shifted vertically due to the differences in delay time at the source points S_1 and S_n. It suffices to take the mean of the two intercept curves to obtain the "isodelay time" curve, associated only with the variations in delay time at the refractor level. This gives the variation of refractor depths h.

These operations can be performed semi-automatically on a computer to obtain the isobath map rapidly. The Gardner method, although quite old, is still often used to map the crystalline basement or to determine the deep structures of the Earth's crust.

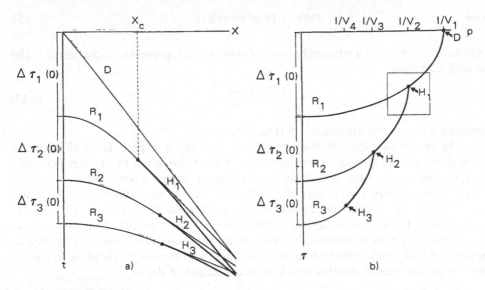

Fig. 6.9 Reflection and headwave traveltimes for three layers. (From J. B. Diebold and P. L. Stoffa, 1981, "The Traveltime Equation, Tau-*p* Mapping and Inversion of Common Midpoint Data", *Geophysics*, Vol. 46, No. 3).

(a) In the (t, x) domain, and **(b)** In the (τ, p) domain. The letters R denote reflections, and the letters H headwaves.

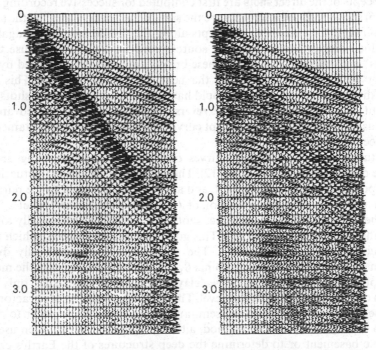

Fig. 6.10 Refraction seismic record before and after filtering by the τ-p transform. Coherent noise extends in the record from 0 to 2.4 s before filtering, and is attenuated by filtering. (*Petty-Ray* Document).

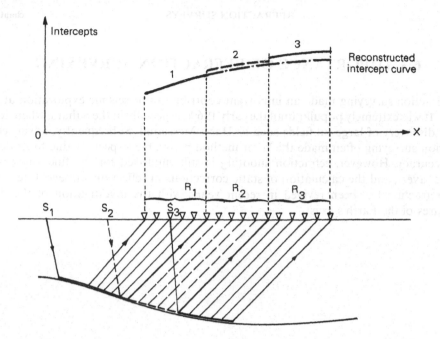

Fig. 6.11 Intercepts 1, 2 and 3 obtained by the Gardner method for three source/detector spreads $S_1 R_1$, $S_2 R_2$ and $S_3 R_3$, and reconstructed intercept curve (direct shots).

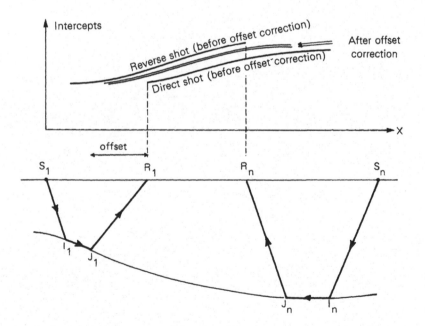

Fig. 6.12. Reconstructed intercepts with direct and reverse shots before and after offset correction.

6.3 IMPORTANCE OF REFRACTION SURVEYING

Refraction surveying made an important contribution to seismic exploration at the outset. It was extremely popular until the early 1960s, especially in the Sahara, where it led to the discovery of large oil fields such as Hassi-Messaoud. Subsequently, advances in reflection surveying often made the latter method preferable, especially due to its detail and accuracy. However, refraction shooting is still employed for the fine analysis of surface layers, and the calculation of static corrections in reflection surveys. Interest in deep refraction has been revived in recent years, with the investigation of the deep structures of the Earth's crust.

CHAPTER

7

---

conclusion

Seismic exploration is an extremely powerful investigation tool, combining great penetration depth with good resolution. Its cost is relatively higher than that of other geophysical methods, but it remains highly competitive due to the accuracy of its diagnosis and the detail of its information.

Tremendous advances have been registered in seismic exploration in the past two decades, thanks to the spectacular growth of data acquisition and processing facilities. Three-dimensional seismic techniques, well seismic profiles, modeling and wavefield inversion algorithms, three-dimensional color displays, interactive processing and interpretation systems have broadened the possibilities of the method.

While the use of seismic methods is still moderate in mining exploration and in hydrology, far behind electrical, magnetic and electromagnetic methods, it is playing an increasingly important role in oceanography and civil engineering. Above all, it is the main tool of petroleum exploration, allowing a safer implantation of exploration wells. Seismic prospecting also contributes to detailed reservoir investigation, and supplements the local information from wireline logging with precise information between the wells.

The problems of the geological interpretation of seismic surveys have not been dealt with in this book. Yet they are the focus of the concerns of exploration engineers. Interpretation is carried out by combined teams of structuralists, stratigraphers and geophysicists. They try to determine the geological structures, to identify the formations traversed by seismic waves, to obtain information about the petrophysical parameters, porosity, fracturation, facies variations, and the presence of hydrocarbons. Interpretation relies on seismic stratigraphy, the reconnaissance of deposition environments and sedimentary bodies, and seismic lithology, the determination of lithological and petrophysical parameters from the analysis of seismic traces.

Interactive graphic systems now make it possible to accelerate the different steps in processing and interpretation. Some expert systems are now being used to determine the implementation parameters. In the future, it is probable that the techniques of artificial intelligence will play an increasingly important role in the acquisition, processing and geological interpretation of seismic surveys.

references

Anstey, N. A., 1981, *Seismic Prospecting Instruments, Vol. 1, Signal Characteristics and Instrument Specifications*, Gebrüder Borntraeger, Berlin.

Backus, M., 1959, "Water Reverberations. Their Nature and Elimination", *Geophysics*, **24**, 233-261.

Baranov, V, Kunetz, G., 1960, "Films synthétiques avec réflexions multiples. Théorie et calcul pratique", *Geophysical Prospecting*, 8, 315-325.

Bath, M., 1974, *Spectral Analysis in Geophysics*, Elsevier Scientific Publishing Co., Amsterdam.

Ben Menahem, A., Singh, S. J., 1981, *Seismic Waves and Sources*, Springer Verlag, Berlin.

Bois, P., Chauveau, J., Grau, G., Lavergne, M., 1960, "Sismogrammes synthétiques. Possibilités, techniques de réalisation et limitations", *Geophysical Prospecting*, 8, 260-314.

Bois, P., Grau, G., Hemon, C., La Porte, M., 1962, "Calcul automatique des sismogrammes synthétiques en ondes planes à l'incidence normale", *Rev. Inst. Franç. du Pétrole*, **17**, 491-522.

Bois, P., La Porte, M., Lavergne, M., Thomas, G., 1972, "Well to Well Seismic Measurements", *Geophysics*, **37**, 471-480.

Brekhovskikh, L. M., 1980, "Waves in Layered Media", *Academic Press*, New York.

Cagniard, L., 1962, *Reflection and Refraction of Progressive Seismic Waves*, translated by E. A. Flin and C. H. Dix, McGraw-Hill, New York.

Cassand, J., Lavergne, M., 1966, *L'Émission sismique par vibrateurs. Le filtrage en sismique*, Éditions Technip, Paris.

Cassand, J., Fail, J. P., Montadert, L., 1970, "Sismique réflexion en eau profonde", *Geophysical Prospecting*, **18**, 600-614.

Chapel, P., 1980, *Géophysique appliquée. Dictionnaire et plan d'Étude*, Masson, Paris.

Claerbout, J., 1976, *Fundamentals of Geophysical Data Processing*, McGraw-Hill, New York.

Coffeen, J. A., 1986, *Seismic Exploration Fundamentals*, Pennwell Publishing Company, Tulsa, Oklahoma, USA.

Cordier, J. P., 1983, *Les Vitesses en sismique réflexion*, Technique et Documentation, Lavoisier, Paris.

Crawford, J. M., Doty, W. E. N., Lee, M. R., 1960, "Continuous Signal Seismograph", *Geophysics*, **25**, 95-105.

Diebold, J. B., Stoffa, P. L., 1981, "The Travel Time Equation, Tau-p Mapping and Inversion of Common Midpoint Data", *Geophysics*, **46**, 238-254.

Dix, C. H., 1952, *Seismic Prospecting for Oil*, Harper and Brothers Press, New York.

Dix, C. H., 1955, "Seismic Velocity from Surface Measurements", *Geophysics*, **20**, 68-86.

Dobrin, M. B., 1976, *Introduction to Geophysical Prospecting*, McGraw-Hill, New York.

Dohr, G., 1985, *Seismic Shear Waves. Handbook of Geophysical Exploration*, Geophysical Press, London.

Evenden, B. S., Stone, D. R., Anstey, N. A., 1971, *Seismic Prospecting Instruments*, Vol. 2, *Instrument Performance and Testing*, Gebrüder Borntraeger, Berlin.

Ewing, W. M., Jardetzky, W. S., Press F., 1957, *Elastic Waves in Layered Media*, McGraw-Hill, New York.

Fitch, A. A., 1979, *Development in Geophysical Exploration Methods*, Applied Science Publishers Ltd., London.

French, W. S., 1974, "Two-dimensional and Three-dimensional Migration of Model Experiment Reflection Profile", *Geophysics*, **39**, 265-277.

Galperin, E. I., 1973, *Vertical Seismic Profiling*, Society of Exploration Geophysicists, Tulsa, Oklahoma, USA.

Gardner, L. W., 1939, "An Area Plan of Mapping Subsurface Structures by Refraction Shooting", *Geophysics*, **4**, 247-259.

Hagedoorn, J. G., 1954, "A Process of Seismic Reflection Interpretation", *Geophysical Prospecting*, **2**, 85-127.

Hardage, B. A., 1983, *Vertical Seismic Profiling. Handbook of Geophysical Exploration*, Geophysical Press, London.

Helbig, K., Treitel, S., 1983/1985, *Seismic Exploration. Handbook of Geophysical Exploration*, Geophysical Press, London.

Ivansson, S., 1985, "A Study of Methods for Tomographic Velocity Estimation in the Presence of Low-Velocity Zones", *Geophysics*, **50**, 969-988.

Kennett, B. L. N., 1983, *Seismic Wave Propagation in Stratified Media*, Cambridge University Press, Cambridge, UK.

Kramer, F. S., Peterson, R. A., Walter, W. C., 1968, *Seismic Energy Sources Handbook*, Bendix United Geophysical, Pasadena, California, USA.

La Porte, M., Lakshmanan, J., Lavergne, M., Willm, C., 1973, "Mesures sismiques par transmission. Application au génie civil", *Geophysical Prospecting*, **21**, 146-158.

Lavergne, M., 1961, "Étude sur modèle ultrasonique du problème des couches minces en sismique-réfraction", *Geophysical Prospecting*, **9**, 60-73.

Lavergne, M., 1970, "Emission by Underwater Explosions", *Geophysics*, **35**, 419-435.

Levin, F. K., 1971, "Apparent Velocity from Dipping Interface Reflections", *Geophysics*, **36**, 510-516.

Love, A. E. H., 1911, *Some Problems of Geodynamics*, Cambridge University Press, Cambridge, UK.

Mechler, P., 1982, *Les méthodes de la géophysique*, Dunod Université, Paris.

McQuillin, R., Bacon, M., Barclay, W., 1979, *Introduction to Seismic Interpretation*, Graham and Trotman, London.

Miller, G. F., Pursey, H., 1954, "The Field and Radiation Impedance of Mechanical Radiators on the Free Surface of a Semi-infinite Isotropic Solid", *Proc. Roy. Soc.*, London, Vol. 223, p. 521.

Millahn, K. O., 1980, "Flözwellenseismik", *Prakla Report* 2, 3/80, pp. 54-65.

Musgrave, A. W., 1967, *Seismic Refraction Prospecting*, Society of Exploration Geophysicists, Tulsa, Oklahoma, USA.

Nelson, Jr., H. R., 1983, *New Technologies in Exploration Geophysics*, Gulf Publishing Co., Houston, Texas, USA.

Newman, P., 1973, "Divergence Effects in a Layered Earth", *Geophysics*, **38**, 481-488.

Peacock, K. L., Treitel , S., 1969, "Predictive Deconvolution. Theory and Practice", *Geophysics*, **34**, 155-169.

Pieuchot, M., 1978, *Les appareillages numériques d'acquisition des données sismiques*, ENSPM, Éditions Technip, Paris.

Pilant, W. L., 1979, *Elastic Waves in the Earth*, Elsevier, Amsterdam.

Polchkov, M. K., Brodov, L., Mironova, L. V., Michon, D., Garotta, R., Layotte, P. C., Coppens, F., 1980, "Combined Use of Longitudinal and Transverse Waves in Reflection Seismics", *Geophysical Prospecting*, **18**, 185-207.

Ricker, N., 1953, "The Form and Laws of Propagation of Seismic Wavelets", *Geophysics*, **18**, 10-40.

Sheriff, R. E., 1973, *Encyclopedic Dictionary of Geophysics*, Society of Exploration Geophysicists, Tulsa, Oklahoma, USA.

Sheriff, R. E., 1980, *Seismic Stratigraphy*, IHRDC, Boston, Massachusetts, USA.

Sheriff, R. E., Geldart, L. P., 1982, *Exploration Seismology*, Cambridge University Press, Cambridge, UK.

Spencer, T. W., Sonnad, J. R., Butler, T. W., 1982, "Seismic Q, Stratigraphy or Dissipation", *Geophysics*, **47**, 16-24.

Taner, M. T., Koehler, F., 1969, "Velocity Spectra. Digital Computer Derivation and Applications of Velocity Functions", *Geophysics*, **34**, 859-881.

Telford, W. M., Geldart, L. P., Sheriff, R. E., Keys, D. A., 1976, *Applied Geophysics*, Cambridge University Press, Cambridge, UK.

Toksoz, M. N., Johnston, D. H., 1981, *Seismic Wave Attenuation*, Geophysics Reprint Series, Society of Exploration Geophysicists, Tulsa, Oklahoma, USA.

Toksoz, M. N., Stewart, R. R., 1984, *Vertical Seismic Profiling. Handbook of Geophysical Exploration*, Geophysical Press, London.

Ville, J., 1948, *Théorie et application de la notion de signal analytique, SSM 3507*, Câbles et Transmissions, Paris.

Waters, K. H., 1978, *Reflection Seismology*, John Wiley and Sons, New York.

White, J. E., 1965, *Seismic Waves, Radiation, Transmission and Attenuation*, McGraw-Hill, New York.

Whitmore, M. G., 1980, "Geophysical Activity in 1979", *Geophysics*, **45**, 1563-1579.

Widess, M. B., 1973, "How Thin is a Thin Bed?", *Geophysics*, **38**, 1176-1180.

Willye, M. R., Gregory, A. R., Gardner, G. H. F., 1958, "An Experimental Investigation of Factors Affecting Elastic Wave Velocities in Porous Media", *Geophysics*, **23**, 459-493.

Wuenschel, P. C., 1960, "Seismogram Synthesis Including Multiples and Transmission Coefficients", *Geophysics*, **25**, 106-129.

Wuenschel, P. C., 1976, "The Vertical Array in Reflection Seismology", *Geophysics*, **41**, 219-232.

Zoeppritz, K., 1919, *Uber Reflexion und Durchgang Seismicher Wellen durch Unstetigkerlsfläschen, über Erdbebenwellen, 7B*, Königlichen Gesellschaft des Wissenschaften, Göttingen, West Germany, pp. 57-84.

subject index

Absorption, 18, 107
Acceleration, 11
Acoustic impedance, 3, 16, 54
Acquisition
  (systems), 65, 81
  of data, 3, 65
AGC (see Automatic Gain Control)
Air
  gun, 70, 71, 72
  wave, 35
Airy phase, 132, 133
Aliasing, 79
Ambiant noise, 41
Amplifier, 79, 80
Amplitude, 81, 106
  spectrum, 44
Analog
  filter, 79
  recording, 79
  to digital converter, 78, 81, 82
Analytical signal, 46
Anchoring system, 137
Angular
  frequency, 18
  relationship, 23
Anisotropic medium, 9
Apparent velocity, 48, 49, 50
Arrival time, 90, 91, 93
Artificial intelligence, 161
Ascending wave, 139, 140, 142
Attenuation, 21, 22, 106
  correction, 106
Autocorrelation, 42, 46, 57, 60, 106, 109, 120
Automatic Gain Control (AGC), 107
Average velocity, 91, 93, 95

Backus deconvolution, 118
Binary gain, 81

Binning, 97, 127
Bit, 81, 106
Blasting, 68
Boundary conditions, 23, 26, 32, 37, 55
Bubble effect, 70, 72
Bulk modulus, 10
Byte, 82

Cable, 83
Calibration on wells, 142
Captain, 103
Causal, 46
Central
  computer, 102
  processing unit, 83
Channel wave, 131
Chief computer, 102
Civil engineering, 131
Client company, 103
CMP (see Common Midpoint)
CMP stacking, 116
Coal mine, 131
Coherence, 114
Collection of traces, 112, 116
Combined mode display, 127
Common Midpoint (CMP), 97, 111, 115, 117,
  151
Common Reflection Point (CRP), 111
Compact sediment, 100
Complex velocity, 19
Compressional wave, 7, 12, 14, 17, 23, 27
Computer, 102
Converted wave, 7, 8, 23, 100, 102
Convolution, 45
Correction (refraction), 155
Correlation, 45

Correlator, 82, 106
CPU (*see* Central Processing Unit)
Critical
    angle, 145, 152, 154
    damping, 76
Crosscorrelation, 45, 57, 82, 106
Cross plot, 127
CRP (*see* Common Reflection Point)
Cutoff frequency, 79

Damping, 76
Data
    acquisition, 65
    processing center, 105
Datum plane, 109
Decca, 86
Deconvolution
    after stacking, 118, 121
    before stacking, 108
Deep reflections, 62
Deformation of pulses, 117, 118
Delay, 153, 157
Demultiplexer, 82
Demultiplexing, 105
Density
    (rock), 13
    log, 53
Depth
    controller, 76
    section, 122
Descending wave, 139, 142
Detection, 63
Detector (seismic), 73, 98
Diffracting point, 123
Diffraction, 38, 39, 40, 123
    hyperbola, 124
    hyperboloid, 127
Digital
    recording, 5, 45, 46, 65, 78, 79
    to analog converter, 78
Digitization in streamer, 68
Digitizing unit, 83
Dilatational
    potential, 12, 25, 31
    stress, 17
    wave (*see* Compressional wave)
Dinoseis, 70
Dip, 91, 123, 154
Dipping
    reflector, 91
    refractor, 153

Dirac pulse, 54, 57, 118
Directivity diagram, 73
Direct shot, 147
Dispersion, 33, 34, 131
Displacement (particle), 9, 15, 16, 23, 31, 55
Display, 126
Dissipation factor, 18, 20
Distortional
    potential, 12, 25, 31
    stress, 17
    wave (*see* Shear wave)
Disturbance time, 11, 13, 15, 42, 65
Dix formula, 95
Doppler
    (sonar), 87, 103
    effect, 87
Downward continuation, 124
Drilling, 68, 102
    crew, 102
Dynamic range, 5, 81
Dynamite charge, 68

Echo, 52, 68
Economic importance, 5
Elasticity, 8
Elastic
    medium, 8
    parameters, 10
Electrical
    impedance, 76
    methods, 5
Electrode, 73
Emission, 65, 68, 70, 72, 135
Energy, 26, 28
ESP (*see* Expanding Spread Profile)
Evanescent wave, 26
Expanding Spread Profile (ESP), 149, 151
Expenditure, 5
Expert systems, 127, 161
Exploration (geophysical), 3, 130
Explosive, 68, 70

Faults, 129, 135
Fermat principle, 20
Field statics, 109
Filtering pattern, 50
Finite difference, 124, 130

Firing time, 83
First arrival, 79
FK method, 124
Flexichoc, 73
Fluid, 21
  wave, 141
Format, 81, 105
Formatter, 81
Fourier transform, 42, 43, 44
Fracture detection, 142
Fracturing, 135
Frequency, 18, 42, 44, 58, 59, 62, 75, 77, 79
  band, 76
  filtering, 44, 78, 79
Fresnel zone, 62

Gain, 78, 79, 107
  recovery, 106
Gardner method, 157
Geological interpretation, 161
Geometric spreading, 22, 106
Geophone, 42, 65, 93, 94
Ghost, 108
Graphic workstation, 127
Ground roll, 33
Group velocity, 34, 37, 132, 133
Gyrocompass, 103

Harmonic component, 19, 42
Heading, 76
Head wave, 151
Helicopter, 102
Helmholtz method, 12
Heterogeneous medium, 20, 22
High resolution, 79
Hilbert transform, 46
Historical background, 4
Homogeneous medium, 8
Hooke's law, 9
Horizontal
  force, 73
  geophone, 101
  layer, 93
  reflector, 90
  refractor, 152

stratification, 92
vibrator, 100
Huyghens principle, 7, 23, 40
Hydrocarbon, 47
Hydrophone, 42, 48, 76

Identification header, 105
Impactor, 70
Impedance matching, 76
Implementation cost, 105
Implosion, 72
Impulse response, 45
Incident angle, 20, 24
Industrial noise, 41
Inelastic medium, 18
Instantaneous
  amplitude, 47
  floating point, 81, 82
  frequency, 47
  phase, 47
Intensity, 22
Interactive workstation, 127, 128
Intercept, 153, 155, 157, 159
  -slowness method, 156
Interface, 3, 23, 28, 65
Interference, 62
Interpretation, 4, 157
Interval velocity, 95
Inversion, 130, 142, 161
Isodelay, 157
Isotropic medium, 8, 10
Iterative method, 133

Kirchhoff method, 124

Laborer, 102
Lamé parameter, 10
Land
  crew, 102
  survey, 65, 68, 83
Laplacian, 11
Lateral resolution, 62
Layer, 53, 60, 92
Least squares method, 108
Lithological
  parameters, 98
  property, 102, 161

Logic unit, 80
Longitudinal wave (see Compressional wave)
Long signal, 57
Lorac, 86
Love wave, 35

Magnet, 75
Magnetic
   compass, 97
   method, 5
   tape, 82, 105
Marine
   crew, 103
   current, 95
   source, 70, 71, 72
   survey, 66, 70, 83
Marthor, 73
Matrix equation, 56
Migration, 123, 127
Mine gallery, 132
Mining
   engineering, 131
   exploration, 5
Minisosie, 70
Mintrop, 4
Mixed representation, 88
Modeling, 130
Modulus, 47
Monitor, 106
Motion (particle), 15, 16, 23, 31, 55
Moveout, 112, 114
Multiple
   coverage, 4, 66, 67, 83, 97, 111, 115
   reflection, 51, 108, 117, 120
   source, 73
Multiplexer, 78, 79
Multiplexing, 78, 79, 83
Multiplicity, 95
Muting, 117
Myriaseis, 84

Navigation, 86, 103, 127
Newman formula, 107
NMO correction, 113, 114, 116, 118, 127
Noise, 41, 62, 68, 116
Normal incidence, 27
Normal moveout (NMO), 112, 114, 116, 118, 127
Nyquist frequency, 79

Observer (laboratory chief), 102, 103
Octave, 79
Offset, 97, 159
Offset VSP, 135
Oil field, 98, 105
Optical fiber, 83
Orchard salt dome, 4

Parity check, 82
Particle velocity, 16, 43
Party chief, 102, 103
Party manager, 102
Pattern, 48, 65, 69
Penetration depth, 130
Periodicity, 50, 79
Permitman, 102
Petrophysical property, 102, 161
Phase
   spectrum, 44, 58
   velocity, 34, 133
Piezo electric sensor, 76
Piling (of layers), 92, 93
Pinchout, 61, 120
Plane wave, 14, 16, 24, 27
Playback, 78, 82
Point source, 73
Poisson's ratio, 11, 33, 98
Porosity, 21
Positioning, 83, 86
Potential, 24, 27, 31
Preamplifier, 76, 78
Predictive deconvolution, 120
Preprocessing, 103
Pressure, 70, 72, 76
Primary reflection, 51
Processing, 105, 127
   center, 105
Propagation
   (wave), 7
   velocity, 13, 18
Pseudo-Rayleigh wave, 33
Pseudo-period, 88
Pulse, 42, 43, 58, 72
P wave (see Compressional wave)

Quadrature signal, 46
Quality factor, 18, 19, 20, 107, 126

Radial displacement, 74
Radio navigation, 103
Radiopositioning, 83
Radio (signal transmission), 83
Ray, 7
Raydist, 86
Rayleigh wave, 31
Raypath, 4, 7, 21
Rebound, 108
Recording, 36, 89
  system, 79
  truck, 83, 102
Reflection
  angle, 24
  coefficient, 23, 51, 54
  hyperbola, 92
  seismogram, 50, 57
  shooting, 68
  surveys, 65
Reflector, 65
Refraction
  shooting, 151
  survey, 145
Reservoir, 102, 133, 142
Residual statics, 109
Resolution, 60
Reverberation, 108, 119
Reverse shot, 147
Rigidity modulus, 10
RMS (Root Mean Square) velocity, 22, 93, 107, 114
Rock matrix, 21
Roll-along, 66
Root Mean Square velocity (see RMS velocity)

Safety, 103
Sampling, 45, 46, 62, 79
Sampling interval, 105
Satellite positioning, 86
Seam, 131
Seismic
  beam, 65
  channel, 79
  crew, 102
  prospecting, 57
  reflection, 3
  refraction, 4, 145
  source, 68, 102
  survey, 3
Semi-infinite medium, 31

Sensitivity matcher, 79
Sensor, 76
Servo cylinder, 68
Shallow reflections, 62
Shear
  modulus, 10
  wave, 7, 12, 17, 23, 29
Ship, 66, 83, 104
Shipboard computer, 95
Ship positioning, 83, 86
Shooter, 83, 102
Short signal, 57
Shot
  command, 83
  sequence, 103
Shunt resistance, 76
SH wave, 16, 23
Signal, 41, 57, 88, 106
  compression, 57, 68
Signature (of a source), 42
Single
  coverage, 65
  source, 73
Slowness, 156
Small refraction, 109
Snell's law, 23
Software, 3
Sonic log, 53
Sonobuoy, 149
Source noise, 41
Soursile, 70
Sparker, 73
Spread, 65, 66, 89, 96
Square grid, 95
Starjet, 72
Static correction, 109, 110, 127
Steam gun, 72
Strain tensor, 9
Streamer, 66, 67, 68, 97
  drift, 97, 127
  noise, 79
  positioning, 76, 95
Stress, 8, 17, 27, 29, 30, 55
  tensor, 8
Supervisor, 103
Surface
  formation, 100
  noise, 50
  wave, 30, 36
SV wave, 15, 24

S wave (*see* Shear wave)
Sweep signal, 57, 106
Switching, 79
Synthetic seismogram, 52

Tabular stratification, 92
Tail buoy, 76
Tangential displacement, 74
Tape unit, 82
Tau-*p* method, 156
TB (Time Break) (*see* Firing time)
Telemetry, 83
Tellurometer, 86
Tensor, 8, 9
Test unit, 80
Theodolite, 102
Three dimensional
    display, 128
    model, 129
    processing, 127
    survey, 95
Time break (*see* Disturbance time)
Time
    domain, 42
    section, 122
Tomography, 133
Topographer, 102
Toran, 86
Tow noise, 79
Trace, 36
Transfer function, 45, 48
Transformer, 76
Transit satellite, 86
Transmission, 23, 131
    coefficient, 23, 27, 51
    losses, 54
    survey, 131
Transversally isotropic medium, 9
Transverse wave (*see* Shear wave)
Tube wave, 141

Unconformity, 120
Unconsolidated formation, 100
Useful frequency band, 76

Vaporchoc, 71

Variable area, 36, 126
Velocity
    analysis, 112, 114, 115, 127
    log, 53
Vertical Seismic Profile (VSP), 133, 143
Vertical
    force, 73
    geophone, 101
    resolution, 60
    vibrator, 68
Vibration truck, 68, 69, 83
Vibrator plate, 68
Vibroseis, 68, 103
Vibroseismic prospecting, 58, 69, 83
Volume dilatation, 10
V_p/V_s Ratio, 100
VSP (*see* Vertical Seismic Profile)

Water gun, 71, 72
Wave equation, 11, 13, 130
    migration, 124
Wave field, 124
Wavefront, 7, 14, 48, 106
Waveguide, 131
Wave length, 18, 34, 42
Wavelet, 42
Wavenumber, 47, 48
Wave surface, 22
Wavetrain, 42
Weathered zone, 33
Weathering shot, 147
Weight drop, 42, 70
Well
    detection, 137
    geophone, 166
    probe, 138
    shooting, 133
Wiggle trace, 126
Wind noise, 79
Workstation, 127
Wyllie formula, 21
WZ (*see* Weathered Zone)

Young's modulus, 10

Zoeppritz equations, 26

ACHEVÉ D'IMPRIMER
EN DÉCEMBRE 1988
PAR L'IMPRIMERIE NOUVELLE
45800 SAINT-JEAN-DE-BRAYE
Nº d'éditeur : 773
Dépôt légal 1988, nº 7440
IMPRIMÉ EN FRANCE